MODERN CHEMISTRY

SECTION REVIEWS

HOLT, RINEHART AND WINSTON

AUSTIN NEW YORK SAN DIEGO CHICAGO TORONTO MONTREAL

AUTHORS

Nicholas D. Tzimopoulos, Ph.D.
Teacher of chemistry and Chairman of the Science Department of the Public Schools of the Tarrytowns, North Tarrytown, New York, and Adjunct Professor of Chemistry at PACE University, New York, New York. He is currently chairman of the Westchester (NY) Section of the American Chemical Society.

New material appearing in this edition of *Modern Chemistry* is the work of Dr. Tzimopoulos.

H. Clark Metcalfe
Formerly teacher of chemistry at Winchester-Thurston School, Pittsburgh, Pennsylvania, and Head of the Science Department, Wilkinsburg Senior High School, Wilkinsburg, Pennsylvania.

John E. Williams
Formerly teacher of chemistry and physics at Newport Harbor High School, Newport Beach, California, and Head of the Science Department, Broad Ripple High School, Indianapolis, Indiana.

Joseph F. Castka
Formerly Assistant Principal for the Supervision of Physical Science, Martin Van Buren High School, New York City, and Adjunct Associate Professor of General Science and Chemistry, C.S. Post College, Long Island University, New York.

Cover: A view of light fuel oil using the technique of phase-contrast photography. In phase-contrast photography, transmitted light is refracted in the same way that a prism breaks white light into a full spectrum of colors. Light fuel oil is one of the many products of fractional distillation of crude oil. Other products in the distillation of crude oil include hydrocarbon polymers, the building blocks of plastics.

Photo: Manfred Kage/Peter Arnold, Inc.

Printed in the United States of America

ISBN 0-03-021878-0
90123456 085 987654321

CONTENTS

PREFACE

MODERN CHEMISTRY SECTION REVIEWS is designed to accompany *MODERN CHEMISTRY.* The reviews are based upon each chapter section in this textbook. Each review generally contains fifteen items. After reading each chapter section in the textbook, these reviews may be completed in order to reinforce your knowledge of material covered in this textbook. Studying these reviews serves as a good survey before a scheduled test. Regular use of the review book will help you establish good work and study habits.

Name ______________________ Class ____________ Date ____________

What Is Chemistry?

Section Review **1.1**

DIRECTIONS: Write on the line at the right of each statement the letter preceding the word or expression that best completes the statement.

1. Chemistry may be least useful in studying (a) matter; (b) synthetic fibers; (c) volcanoes; (d) medicine. ______ 1
2. The science most concerned with studying the composition of materials is (a) biology; (b) chemistry; (c) microbiology; (d) astronomy. ______ 2
3. One of the most important tools in chemistry is the (a) beaker; (b) periodic table; (c) scale; (d) magnifying glass. ______ 3
4. A manager of a pharmaceutical company, in addition to being trained in management, would most likely be trained in (a) engineering; (b) chemistry; (c) physics; (d) geology. ______ 4
5. To determine the causes and effects of chemical pollution in a large body of water would require chemists who are trained in (a) physics and chemistry; (b) chemistry, geology, and physics; (c) chemistry and biology; (d) chemistry, biology, and aeronautics. ______ 5
6. Many items of clothing would not exist without the technology chemistry has developed in the area of (a) aramid fibers; (b) synthetic fibers; (c) recycling; (d) fiberglass. ______ 6

DIRECTIONS: Define five major branches of chemistry.

7. (1) ______________________ 7
8. (2) ______________________ 8
9. (3) ______________________ 9
10. (4) ______________________ 10
11. (5) ______________________ 11

DIRECTIONS: Describe in detail the four steps of the scientific method.

12. Step 1 ______________________ 12
13. Step 2 ______________________ 13
14. Step 3 ______________________ 14
15. Step 4 ______________________ 15

Name ____________________ Class __________ Date __________

Matter and Energy

Section Review 1.2

DIRECTIONS: Write on the line at the right of each statement the letter preceding the word or expression that best completes the statement.

1. The weight of an object is least (a) at sea level; (b) at a high altitude; (c) at the bottom of a deep mine shaft; (d) in a sunken vessel. ______ 1
2. Inertia is most closely related to a body's (a) weight; (b) mass; (c) density; (d) volume. ______ 2
3. The chemical energy that substances possess because of their composition and structure is a kind of (a) kinetic energy; (b) nuclear energy; (c) potential energy; (d) electric energy. ______ 3
4. The ability to cause change is called (a) energy; (b) chemistry; (c) electricity; (d) velocity. ______ 4
5. The particles in a solid are (a) packed closely together; (b) very far apart; (c) constantly in motion; (d) able to slide past each other. ______ 5
6. A change in temperature causes the most change in volume of a (a) solid; (b) liquid; (c) gas; (d) element. ______ 6
7. An experiment which determines the density of water is investigating (a) a chemical change; (b) a physical property; (c) the potential energy of water; (d) a chemical property. ______ 7
8. A chemical property of oxygen is that it (a) is the most abundant element in the earth's crust; (b) combines with hydrogen to form water; (c) boils at -183 degrees Celsius; (d) has a density of 1.43 g/L. ______ 8
9. A dark gray solid is heated in a closed container and purple vapors are given off. These purple vapors condense to a dark gray powder at the top of the container. This is an example of (a) a chemical change; (b) a physical change; (c) a boiling point determination; (d) a combustion. ______ 9
10. Road crews spread salt in the winter because the salt, as it dissolves, releases enough heat to melt ice. This heat indicates (a) an endothermic reaction; (b) that gas is being produced; (c) an exothermic reaction; (d) the formation of a precipitate. ______ 10
11. Chemical potential energy stored in reactants is (a) released in an exothermic reaction; (b) released in an endothermic reaction; (c) converted to kinetic energy in an endothermic reaction; (d) converted to electrical energy in a chemical reaction. ______ 11

DIRECTIONS: Write the answers to the following on the lines provided. Where appropriate, make complete statements.

12. Define and give examples of physical properties. ______________________________

____

______________________________ 12

13. Describe the gaseous state in terms of particles. ______________________________

____

______________________________ 13

14. The law of conservation of energy states ______________________________

____

______________________________. 14

15. The law of conservation of matter states ______________________________

____

______________________________. 15

Name ______________________ Class __________ Date __________

Classification of Matter

Section Review 1.3

DIRECTIONS: Write on the line at the right of each statement the letter preceding the word or expression that best completes the statement.

1. All of the following are mixtures with the exception of (a) brass; (b) air; (c) baking soda; (d) concrete. ______ 1
2. Water composed of 11.2% hydrogen and 88.8% oxygen by mass is (a) a mixture; (b) a solution; (c) heterogeneous; (d) a pure substance. ______ 2
3. Pure substances, solutions and elements (a) all contain phases; (b) are mixtures; (c) are heterogeneous; (d) are homogeneous. ______ 3
4. Pure substances can be (a) solutions; (b) heterogeneous; (c) elements; (d) mixtures. ______ 4
5. A compound composed of three elements is (a) salt; (b) water; (c) sugar; (d) carbon dioxide. ______ 5

DIRECTIONS: Complete the following table by checking either homogeneous or heterogeneous, depending on the type of substance.

	Substance	Homogeneous	Heterogeneous	
6.	iron ore			6
7.	quartz			7
8.	granite			8
9.	pancake syrup			9
10.	vegetable soup			10
11.	salt			11
12.	water			12
13.	nitrogen			13

DIRECTIONS: Write the answers to the following on the lines provided.

14. State the law of definite composition. ______________________________

____

____

______________________________ 14

15. Identify and give examples of the two primary types of homogeneous matter. ______________________________

____

____

____

______________________________ 15

Name ______________________ Class ____________ Date ____________

The Chemical Elements

Section Review 1.4

DIRECTIONS: Complete each of the following statements, forming accurate sentences.

1. A horizontal row of blocks in the periodic table is called a(n) ____________________. 1
2. The symbol for the element in Period 2, Family 3, is ____________________. 2
3. An element that is a good conductor of heat and electricity is called a(n) ____________________. 3
4. An element that is a poor conductor of heat and electricity is called a(n) ____________________. 4

DIRECTIONS: Complete the following table by filling in the spaces with correct names and symbols.

	Common Elements and Their Symbols		
	Name	**Symbol**	
5.	aluminum		5
6.		Ca	6
7.		Mn	7
8.	nickel		8
9.	phosphorus		9
10.	cobalt		10
11.		Si	11
12.		H	12

DIRECTIONS: Write the answers to the following on the lines provided.

13. Define metalloid and give an example of where metalloids are used in everyday life. ____________________ 13

14. Name three characteristics of nonmetals. ____________________ 14

15. Name three characteristics of metals. ____________________ 15

Name ______________________ Class __________ Date __________

Section Review 2.1

Units of Measurement

DIRECTIONS: Write on the line at the right of each statement the number that makes the statement an equality when substituted for the corresponding number.

1. 100 mL = __(1)__ dL ____________ 1
2. 10^{-2} m = __(2)__ mm ____________ 2
3. 0.25 g = __(3)__ mg ____________ 3
4. 0.06 cm = __(4)__ mm ____________ 4
5. 2.04 L = __(5)__ mL ____________ 5

DIRECTIONS: Write on the line at the right of each statement the letter preceding the word or expression that best completes the statement.

6. The metric prefix deci- (symbol d) represents the factor (a) 10^{-1}; (b) 10^{-2}; (c) 10^{-3}; (d) 1^{-1}. ________ 6
7. The internationally accepted units for the fundamental quantities of length and mass are (a) centimeter and gram; (b) meter and gram; (c) centimeter and kilogram; (d) meter and kilogram. ________ 7
8. The metric unit for length that is closest to the thickness of a dime is (a) micrometer; (b) millimeter; (c) centimeter; (d) decimeter. ________ 8
9. If the density of a substance is multiplied by the volume, and the product obtained is divided by the number of grams per mole, the unit for the answer that results will be (a) liters per mole; (b) mole; (c) grams; (d) moles per gram. ________ 9
10. Starting with 364 days per year, you can calculate the number of seconds in a year by multiplying the number of hours per day and (a) dividing by min/hr and multiplying by sec/min; (b) dividing by both min/hr and sec/min; (c) multiplying by both min/hr and sec/min; (d) multiplying by min/hr and dividing by sec/min. ________ 10
11. The quantity expressed by m^3 is (a) length; (b) mass; (c) volume; (d) density. ________ 11
12. The liter is specifically defined as (a) 1000 mL; (b) 1 dm^3; (c) 10^3 cm^3; (d) 1.06 qt. ________ 12

DIRECTIONS: Write the answer to questions 13–15 on the line to the right, and show your work in the space provided.

13. Aluminum has a density of 2.70 g/cm^3. Calculate the mass of a solid piece of aluminum whose volume is 1.50 cm^3. ____________ 13

14. A 5.00 sample of gold has a mass of 96.5 grams. Calculate the density of gold. ____________ 14

15. Calculate the density of 37.72 g of matter whose volume is 6.8 cm^3. ____________ 15

Name ______________________ Class ____________ Date ____________

Heat and Temperature

Section Review 2.2

DIRECTIONS: Write on the line at the right of each statement the number that makes the statement an equality when substituted for the corresponding number.

1. 30 °C = __(1)__ K ______ 1
2. −25 °C = __(2)__ K ______ 2
3. 190 K = __(3)__ °C ______ 3
4. 300 K = __(4)__ °C ______ 4
5. 6500 cal = __(5)__ kj ______ 5
6. 1.62 kj = __(6)__ kcal ______ 6

DIRECTIONS: Write on the line at the right of each statement the letter preceding the word or expression that best completes the statement.

7. Determining whether an object feels hot or cold to the touch is a way of measuring its (a) temperature; (b) heat; (c) density; (d) none of the above. ______ 7
8. The measurement using a thermometer is based on the (a) type of thermometer liquid; (b) quantity of thermometer liquid; (c) uniform expansion or contraction of the thermometer liquid; (d) thermometer liquid's color. ______ 8
9. Water freezes at (a) 273.15 K; (b) 100 °C; (c) 0 K; (d) 373.15 K. ______ 9
10. The equation used to convert from the Celsius to Kelvin scale is (a) T(K) = t(°C) + 100; (b) T(K) = t(°C) + 273.15; (c) T(C) = t(K) + 100; (d) T(°C) = t (K) + 273.15. ______ 10
11. If two systems do not have heat flowing between them, they have the same (a) density; (b) thermodynamic properties; (c) specific heat; (d) temperature. ______ 11
12. One calorie is the same amount of heat as 4.184 (a) joules; (b) Calories; (c) kilojoules; (d) kilocalories. ______ 12

DIRECTIONS: Write the answer to questions 13–15 on the line to the right, and show your work in the space provided.

13. A 4.0 g sample of iron was heated from 0 °C to 20 °C, and was found to have absorbed 35.2 J of heat. Calculate the specific heat of this piece of iron. ______ 13
14. Calculate how much heat a copper sample will gain if its specific heat is 0.384 J/g°C, its mass is 8 g, and it is heated from 10 °C to 40 °C. ______ 14
15. Determine the specific heat of a material if a 6 g sample absorbs 50 J as it is heated from 30 °C to 50 °C. ______ 15

Name ______________________ Class ____________ Date ____________

Using Scientific Measurements

Section Review 2.3

DIRECTIONS: Write the number of significant figures in each measurement on the line to the right.

1. 0.000305 kg ______ 1
2. 32.2 m ______ 2
3. 240.020 km ______ 3
4. 30.00 dL ______ 4
5. 210 g ______ 5

DIRECTIONS: Write the number that makes each statement an equality on the line to the right. Round your answer to the correct number of significant figures, and show your work in the space provided.

6. 1.27 cm × 1.3 cm × 2.5 cm = __(6)__ cm^3 ________ 6

7. 5.7 m divided by 2 m = __(7)__ m ________ 7

8. 214.53 km + 32 km = __(8)__ km ________ 8

9. 12 cm × 5.7 cm = __(9)__ cm^2 ________ 9

10. 0.13 g + 1.7 g + 0.562 g = __(10)__ g ________ 10

DIRECTIONS: Write each number expressed in scientific notation on the line to the right.

11. 0.000065 cm = ______________________ cm 11
12. 0.0930 m = ______________________ m 12
13. 5900 km = ______________________ km 13

DIRECTIONS: Write on the line at the right of each statement the letter preceding the word or expression that best completes the statement.

14. Uncertainty in scientific measurement due to poor precision can result from (a) the standard being too strict; (b) human error; (c) limitations of the measuring device; (d) both b and c. ______ 14

15. A chemist who frequently carries out a complex experiment is likely to have a great deal of (a) accuracy, but not any precision; (b) accuracy; (c) precision; (d) precision, but not any accuracy. ______ 15

Name ______________________ Class ____________ Date ____________

Solving Quantitative Problems

Section Review 2.4

DIRECTIONS: Use the four-step method to solve questions 1–4. Write your answer on the line at the right, and show your work in the space provided.

1. What is the volume of a cube that measures 2.5 cm on a side? __________ 1

2. If one inch equals 2.54 centimeters, how many centimeters equal one yard? __________ 2

3. How many minutes are in two weeks? __________ 3

4. You expect to drive an average speed of 79 km/hr on a four hour trip. How far will you travel? __________ 4

DIRECTIONS: Write the answer to questions 5–10 on the line to the right, and show your work in the space provided. Questions 5–10 refer to the equation for calculating density, or $D = m/v$.

5. The equation for density represents what type of proportion? __________ 5

6. In the formula for density, which variables have a direct proportionality? __________ 6

7. In the formula for density, which measurement is a constant? __________ 7

8. When mass of a substance equals 1 g and its volume is 1.5 cm^3, what is its density? __________ 8

9. What is the volume of a 3 gram sample of a substance whose density equals 1.5 g/cm^3? __________ 9

10. This equation will graph as what type of line? __________ 10

DIRECTIONS: List the steps in the four-step method for solving quantitative problems.

11. Step (1) ______________________________ 11

12. Step (2) ______________________________ 12

13. Step (3) ______________________________ 13

14. Step (4) ______________________________ 14

15. Draw a rough sketch of a graph representing a directly proportional relationship between the variables on the x and y axis. 15

Name ________________________ Class ____________ Date ____________

The Atom: From Philosophical Idea to Scientific Theory

Section Review 3.1

DIRECTIONS: Write on the line at the right of each statement the letter preceding the word or expression that best completes the statement.

1. The schoolmaster who studied atoms and proposed an atomic theory was (a) John Dalton; (b) Jons Berzilius; (c) Johann Dobereiner; (d) Dmitri Mendeleev. ______ 1
2. According to Dalton's atomic theory, atoms (a) are destroyed in chemical reactions; (b) can be subdivided; (c) of a particular element are identical in size, mass, and other properties; (d) of different elements cannot combine. ______ 2
3. One part of Dalton's atomic theory that has been modified is the idea that (a) all matter is composed of atoms; (b) atoms of different elements have different properties and masses; (c) atoms can combine in chemical reactions; (d) atoms cannot be subdivided. ______ 3
4. Dalton's atomic theory successfully explained the law of (a) whole-number ratios; (b) definite proportions; (c) conservation of mass; (d) conservation of energy. ______ 4
5. The law of definite composition (a) contradicted Dalton's atomic theory; (b) was explained by Dalton's atomic theory; (c) replaced the law of conservation of mass; (d) assumes that atoms of all elements are identical. ______ 5
6. The fact that lead forms two oxides of different formulas, PbO and PbO_2, is an example of the (a) periodic law; (b) law of multiple proportions; (c) atomic law; (d) law of conservation of mass. ______ 6

DIRECTIONS: Write on the line at the right of each statement the word or expression that best completes the meaning when substituted for the corresponding number.

7. Water, H_2O, has a mass ratio of oxygen to hydrogen of 8:1. Hydrogen peroxide, H_2O_2, has a mass ratio of oxygen to hydrogen of __(7)__ . ________ 7
8. If 3 grams of element C combine with 8 grams of element D to form a compound CD, __(8)__ grams of D are needed to form compound CD_2. ________ 8
9. Evidence in support of the law of __(9)__ is that in oxides of nitrogen, such as N_2O, NO, NO_2, and N_2O_3, atoms combine in small whole-number ratios. ________ 9
10. __(10)__ is the person credited with being the first to recognize that the relative number of atoms that combine are proportional to the masses that combine. ________ 10
11. An example of the law of __(11)__ is the fact that the mass ratio of two elements in a compound is constant. ________ 11
12. If atoms of element D weigh three mass units and atoms of element E weigh five mass units, a chemical compound composed of one atom each of D and E will weigh __(12)__ mass units. ________ 12
13. If 2 grams of element A combine with 10 grams of element B, then 12 grams of element A will combine with __(13)__ grams of element B. ________ 13

DIRECTIONS: Write the answers to the following on the lines provided.

14. State the law of multiple proportions. ____

____

____

____ 14

15. State the two main ideas of Dalton's atomic theory that have remain unchanged since he first proposed his theory. ____

____

____ 15

Name ______________________ Class ____________ Date ____________

The Structure of the Atom

Section Review 3.2

DIRECTIONS: Write on the line at the right of each statement the letter preceding the word or expression that best completes the statement.

1. In early experiments on electricity and matter, electrical current was passed through a glass tube containing (a) water; (b) gas under high pressure; (c) liquid oxygen; (d) gas under low pressure. ______ 1
2. As a result of the movement of a paddle wheel placed between the electrodes in a glass tube through which electrical current passed, scientists concluded that (a) a magnetic field was produced; (b) particles were passing from the cathode to the anode; (c) there was gas in the tube; (d) atoms were indivisible. ______ 2
3. Since most particles fired at metal foil pass through the foil, it may be concluded that (a) atoms are mostly empty space; (b) atoms contain no charged particles; (c) electrons form the nucleus; (d) atoms are indivisible. ______ 3
4. Since a few positively-charged particles bounce back from metal foil, it may be concluded that (a) an atom is indivisible; (b) electrons make up the center of atoms; (c) an atom carries a positive charge; (d) an atom contains a small, dense, positively-charged central region. ______ 4
5. The nucleus of an atom has all of the following characteristics EXCEPT that it (a) is positively charged; (b) is very dense; (c) contains nearly all of the atom's mass; (d) contains nearly all of the atom's volume. ______ 5
6. An atom is electrically neutral because (a) neutrons balance the protons and electrons; (b) nuclear forces equalize the charges; (c) the number of protons and electrons is equal; (d) the number of protons and neutrons is equal. ______ 6
7. The most common form of hydrogen has (a) no neutrons; (b) 1 neutron; (c) 2 neutrons; (d) 3 neutrons. ______ 7

DIRECTIONS: Complete the following statements, forming accurate sentences.

8. The only radioactive form of hydrogen is ______________________. 8
9. The mass number of deuterium is ______________________. 9
10. The name of the scientist who showed the existence of the nucleus by bombarding the gold foil with positively charged particles and noting that some were deflected was ______________________. 10
11. In the glass tubes used to study the nature of matter, electrical current passed from the negative electrode that is called the ______________________. 11
12. The smallest unit of an element that can exist either alone or in combination with atoms of the same or different elements is the ______________________. 12
13. A positively charged particle with a mass of 1.673×10^{-24} g is a(n) ______________________. 13
14. A nuclear particle that has no electrical charge, is called a(n) ______________________. 14
15. Isotopes are atoms of the same element that have different ______________________. 15

Name ______________________ Class ____________ Date ____________

Weighing and Counting Atoms

Section Review 3.3

DIRECTIONS: Write on the line at the right of each statement the letter preceding the word or expression that best completes the statement.

1. The atomic number of oxygen, 8, indicates that there are eight (a) protons in the nucleus of an oxygen atom; (b) oxygen nuclides; (c) neutrons outside the oxygen atom's nucleus; (d) energy levels moving about each nucleus. _______ 1
2. The number that is the total number of protons and neutrons in the nucleus of an atom is the (a) atomic number; (b) Avogadro number; (c) mass number; (d) number of neutrons. _______ 2
3. Isotopes of a particular element differ in (a) the number of neutrons in the nucleus; (b) atomic number; (c) the number of electrons in their outermost energy level; (d) the total number of electrons. _______ 3
4. In determining atomic mass units, the standard is a(n) (a) C–12 atom; (b) C–14 atom; (c) H–1 atom; (d) O–16 atom. _______ 4
5. The chemical shorthand for atomic mass units is (a) amu; (b) mu; (c) a; (d) *u*. _______ 5
6. The average atomic mass of an element is an average of the atomic masses of the (a) naturally-occurring isotopes; (b) two most abundant isotopes; (c) nonradioactive isotopes; (d) artificial isotopes. _______ 6
7. The number of atoms in one mole of carbon is (a) 6.022×10^{22}; (b) 6.022×10^{23}; (c) 5.022×10^{22}; (d) 5.022×10^{23}. _______ 7
8. The number of atoms in a mole of any substance is called (a) the atomic number; (b) Avogadro's number; (c) the mass number; (d) the gram-atomic number. _______ 8
9. As the atomic masses of the elements in the periodic table increase, the number of atoms in one mole of each element (a) decreases; (b) increases; (c) remains the same; (d) becomes a negative number. _______ 9

DIRECTIONS: Write the answer to questions 10–15 on the line to the right, and show your work in the space provided.

10. Oxygen has isotopes: O–16, O–17, and O–18. Naturally occurring oxygen consists of 99.759% of oxygen–16, atomic mass 15.99491 *u*. What is the average atomic mass of the mixture of isotopes? __________ 10
11. How many neutrons does Zn–66 (atomic number 30) have? __________ 11
12. Ag–109 has 62 neutrons. How many electrons does the neutral atom have? __________ 12
13. An atom of chlorine has an atomic number of 17 and a mass number of 35. How many protons, electrons, and neutrons does it have? __________ 13
14. Neon has 3 natural isotopes: Ne–20 (90.51%, 19.99244 *u*), Ne–21 (0.27%, 20.99395 *u*), and Ne–22 (9.22%, 21.99138 *u*). What is neon's average atomic mass? __________ 14
15. What is the mass of two moles of oxygen atoms if the atomic mass of oxygen is 16? __________ 15

Name ______________________ Class ____________ Date ____________

Refinements of the Atomic Model

Section Review 4.1

DIRECTIONS: Write on the line at the right of each statement the letter preceding the word or expression that best completes the statement.

1. Electromagnetic radiation has some properties of particles when it (a) travels through space; (b) is transferred to matter; (c) interacts with photons; (d) interacts with other radiations. ________ 1
2. The wave model of light was not able to explain (a) light's frequency; (b) the continuous spectrum; (c) interference; (d) the photoelectric effect. ________ 2
3. In wave motion, the product of frequency and wavelength is equal to (a) the number of waves passing a given point in a second; (b) the speed of the wave; (c) the distance between successive wave crests; (d) the time for one full wave to pass a given point. ________ 3
4. The common characteristic shown by X rays, visible light, infrared radiation, and radio waves is that they all have the same (a) energy; (b) wavelength; (c) speed; (d) frequency. ________ 4
5. Red light has a longer wavelength than blue light. Compared to the blue line on the hydrogen spectrum, the red line would represent (a) higher energy and lower frequency; (b) higher energy and higher frequency; (c) lower energy and higher frequency; (d) lower energy and lower frequency. ________ 5
6. A line spectrum is produced when an electron moves from one energy level (a) to a higher energy level; (b) to a lower energy level; (c) into the nucleus; (d) to another position in that same sublevel. ________ 6
7. The drop of an electron from a high energy level to the ground state in a hydrogen atom would be most closely associated with (a) long wavelength radiation; (b) low frequency radiation; (c) infrared radiation; (d) high frequency radiation. ________ 7
8. The change of an atom from excited state to ground state always involves (a) absorption of energy; (b) emission of electromagnetic radiation; (c) release of visible light; (d) an increase in electron energy. ________ 8
9. An orbital may be defined as (a) the most stable state of an atom; (b) the circular path followed by an electron around the nucleus; (c) the positively charged central part of an atom; (d) a highly probable location of an electron within the atom. ________ 9
10. The quantum model of the atom locates the electron (a) at a specific distance from the nucleus; (b) in a definite path around the nucleus; (c) within a region of high probability; (d) at any distance from the nucleus. ________ 10
11. The size and shape of an electron cloud is most closely related to the electron's (a) charge; (b) mass; (c) spin; (d) energy. ________ 11

DIRECTIONS: Complete the following statements, forming accurate sentences.

12. A quantum of electromagnetic energy is called a(n) ______________________. 12
13. The spectral lines of hydrogen that occur in the ultraviolet region of the electromagnetic spectrum are called the ______________________. 13
14. An optical instrument that separates light entering it into component wavelengths is a(n) ______________________. 14
15. The lowest total energy of the electron in a hydrogen atom occurs when the electron is in the state called the ______________________. 15

Name ____________________ Class __________ Date __________

Quantum Numbers and Atomic Orbitals

Section Review 4.2

DIRECTIONS: Write on the line at the right of each statement the letter preceding the word or expression that best completes the statement.

1. How many quantum numbers are used to describe the energy state of an electron in an atom? (a) 1 (b) 2 (c) 3 (d) 4 ______ 1
2. A spherical electron cloud surrounding an atomic nucleus would best represent (a) an *s* orbital; (b) a p_x orbital; (c) a combination of p_x and p_y orbitals; (d) a combination of an *s* and a p_x orbital. ______ 2
3. The letter designations for the first four orbital quantum numbers with the number of spatial positions available for orbitals at each sublevel are (a) *s*:1, *p*:3, *d*:10, and *f*:14; (b) *s*:1, *p*:3, *d*:5, and *f*:7; (c) *s*:2, *p*:6, *d*:10, and *f*:14; (d) *s*:1, *p*:2, *d*:3, and *f*:4. ______ 3
4. The number of possible orbital shapes for the third energy level is (a) 1; (b) 2; (c) 3; (d) 4. ______ 4
5. The maximum number of electrons that can occupy the *s* orbitals at each energy level is (a) two, if they have opposite spins; (b) two, if they have the same spin; (c) one; (d) no more than eight. ______ 5
6. The spin quantum number indicates that the number of possible orientations for an electron in an orbital is (a) 1; (b) 2; (c) 3; (d) 5. ______ 6
7. The values +½ and −½ specify an electron's (a) charge; (b) main energy level; (c) speed; (d) possible orientation in an orbital. ______ 7
8. When n represents the principal quantum number of an energy level, the number of electrons per energy level is (a) n; (b) 2n; (c) n^2; (d) $2n^2$. ______ 8
9. At $n = 1$, the total number of electrons that could be found is (a) 1; (b) 2; (c) 6; (d) 18. ______ 9

DIRECTIONS: Write on the line at the right of each statement the word or expression that best completes the meaning when substituted for the corresponding number.

10. The quantum number that indicates the position of an orbital about the three axes in space is the __(10)__ quantum number. ______ 10
11. The __(11)__ orbitals are dumbbell-shaped and directed along the x, y, and z axes. ______ 11
12. The number of different sublevels within each energy level of an atom is equal to the value of the __(12)__ quantum number. ______ 12
13. There are __(13)__ orbitals for the *d* sublevel. ______ 13
14. __(14)__ electrons are needed to completely fill the fourth energy level. ______ 14
15. A particular main energy level can hold 18 electrons. In this case, *n* equals __(15)__ . ______ 15

Name ______________________ Class __________ Date __________

Electron Configurations

Section Review **4.3**

DIRECTIONS: Write on the line at the right of each statement the letter preceding the word or expression that best completes the statement.

1. The correct sequence in ascending energies of atomic sublevels is (a) 4*d*, 5*s*, 5*p*, 6*s*; (b) 5*s*, 4*d*, 5*p*, 6*s*; (c) 5*s*, 4*d*, 6*s*, 5*p*; (d) 5*s*, 5*p*, 4*d*, 6*s*. ______ 1
2. The statement that an electron occupies the lowest energy orbital that can receive it is (a) Hund's rule; (b) the Aufbau principle; (c) Bohr's law; (d) the Pauli exclusion principle. ______ 2
3. In the correct electron-dot notation for the phosphorus atom (atomic number 15), the symbol P is surrounded by (a) two pairs of dots and a single dot; (b) three pairs of dots and a single dot; (c) one pair of dots and three single dots; (d) two pairs of dots. ______ 3
4. The electron-configuration notation for scandium (atomic number 21) would show the three highest energy electrons to have the notation (a) $3d^1 4s^2$; (b) $4s^2$; (c) $3d^3$; (d) $4s^2 4p^1$. ______ 4
5. The element with the electron-configuration notation $1s^2 2s^2 2p^6 3s^2 3p^6$ is (a) Mg (z=12); (b) P (z=15); (c) S (z=16); (d) Si (z=14). ______ 5
6. In the correct electron-dot notation for sulfur (atomic number 16), the symbol S is surrounded by (a) three pairs of dots; (b) two pairs of dots and two single dots; (c) four single dots; (d) two pairs of dots. ______ 6
7. If the highest main energy level of an atom has the *s* and *p* orbitals filled with electrons, it is said to have a(n) (a) electron pair; (b) octet; (c) ellipsoid; (d) circle. ______ 7
8. The noble gas configuration is an outer main energy level fully occupied by (a) 32 electrons; (b) 8 electrons; (c) 2 electrons; (d) 64 electrons. ______ 8

DIRECTIONS: Complete the following statements, forming accurate sentences.

9. The atomic sublevel with the next highest energy after 4*p* is ______________________. 9
10. "Orbitals of equal energy are each occupied by one electron before any is occupied by a second electron" is a statement of ______________________. 10
11. The electron-dot notation for an element in the third series is represented by a symbol surrounded by a pair of dots and a single dot. The complete electron configuration for this element is $1s^2 2s^2 2p^6$ ______________. 11
12. The electron configuration for the carbon atom (C) is $1s^2 2s^2 2p^2$. The atomic number of carbon is __________. 12
13. The electron-configuration notation for the element with atomic number 11 is ______________________. 13
14. The number of electrons in the highest energy level of the argon atom (atomic number 18) is __________. 14
15. An element with eight electrons in its outermost main energy level is called a(n) ______________________. 15

Name ______________________________ Class ____________ Date ____________

History of the Periodic Table

Section Review 5.1

DIRECTIONS: Write on the line at the right of each statement the letter preceding the word or expression that best completes the statement.

1. Mendeleev attempted to organize the chemical elements based on their (a) symbols; (b) properties; (c) atomic numbers; (d) electron configurations. _______ 1
2. Mendeleev noticed that properties of elements appeared at regular intervals when the elements were arranged in order of increasing (a) atomic number; (b) density; (c) reactivity; (d) atomic mass. _______ 2
3. The most useful source of general information about the elements associated with chemistry is a (a) calculator; (b) table of metric equivalents; (c) periodic table; (d) table of isotopes. _______ 3
4. Elements in a group, or column, in the periodic table can be expected to have similar (a) atomic masses; (b) atomic numbers; (c) numbers of neutrons; (d) properties. _______ 4
5. The radioactive elements with atomic numbers from 90 to 103 in the periodic table are referred to as (a) noble gases; (b) lanthanide elements; (c) actinide elements; (d) rare earth elements. _______ 5
6. Argon, krypton, and xenon are all (a) alkaline earth metals; (b) noble gases; (c) actinides; (d) lanthanides. _______ 6
7. Lithium, the first element in Group 1, has an atomic number of 3. The second element in this group will have an atomic number of (a) 4; (b) 10; (c) 11; (d) 18. _______ 7
8. Krypton, atomic number 36, is the fourth element in Group 18. What is the atomic number of xenon, the fifth element in Group 18? (a) 54 (b) 68 (c) 72 (d) 90 _______ 8

DIRECTIONS: Write on the line at the right of each statement the word or expression that best completes the meaning when substituted for the corresponding number.

9. Mendeleev left spaces in his periodic table and predicted the existence and ___(9)___ of several elements. __________ 9
10. The first successful ___(10)___ was developed by Mendeleev. __________ 10
11. "The physical and chemical properties of elements are periodic functions of their atomic ___(11)___" is the Periodic Law. __________ 11
12. The work of a person named ___(12)___ led to a periodic table based on increasing atomic number. __________ 12
13. A(n) ___(13)___ is a horizontal row of blocks in the periodic table. __________ 13
14. The ___(14)___ are the elements whose discovery added an entirely new row to Mendeleev's periodic table. __________ 14
15. For Groups 1, 2, and 18, the atomic numbers of the fourth element in the group is ___(15)___ more than the preceding element. __________ 15

Name ____________ Class ________ Date ________

Electron Configuration and the Periodic Table

Section Review 5.2

DIRECTIONS: Write on the line at the right of each statement the letter preceding the word or expression that best completes the statement.

1. How many elements will there be in a period that involves the filling of *s* and *p* sublevels only? (a) 2 (b) 8 (c) 18 (d) 32 ______ 1
2. The length of each period in the periodic table is determined by the (a) atomic masses of the elements; (b) atomic numbers of the elements; (c) sublevels being filled with electrons; (d) number of isotopes of each element. ______ 2
3. Elements to the right-hand side of the periodic table, or *p*-block elements, tend to have properties most closely associated with (a) gases; (b) nonmetals; (c) metals; (d) metalloids. ______ 3
4. Elements in which the *d*-sublevel is being filled have the properties of (a) metals; (b) nonmetals; (c) metalloids; (d) gases. ______ 4
5. For Groups 13–18, the total number of electrons in the highest occupied level equals the group number (a) plus 1; (b) minus 1; (c) plus 5; (d) minus 10. ______ 5
6. Where *n* stands for the highest occupied energy level, the outer configuration for all Group 1 elements is represented as (a) ns^1; (b) $2n$; (c) $n - s$; (d) np^1. ______ 6
7. In nature, the alkali metals exist as (a) elements; (b) compounds; (c) complex ions; (d) gases. ______ 7
8. Compared to the alkali metals, the alkaline earth metals (a) are less reactive; (b) have lower melting points; (c) are less dense; (d) combine more readily with nonmetals. ______ 8

DIRECTIONS: Complete the following statements, forming accurate sentences.

9. Aluminum, atomic number 13, has the electron configuration [Ne] $3s^23p^1$. Aluminum is in Period ________. 9
10. Cesium, atomic number 55, has the electron configuration [Xe] $6s^1$. Cesium is in Period ________. 10
11. Neutral atoms with an s^2p^6 electron configuration in the highest energy level are named the ________________. 11
12. The elements that border the zigzag line in the periodic table are called the ________. 12
13. The elements characterized by the electron-configuration notation of s^2p^5 in the highest occupied energy level belong to Group ________. 13
14. Strontium's highest occupied energy level is $4s^2$. Strontium belongs to Group ________. 14
15. The most active member of the halogen group, with atomic number 9, is ________. 15

Name ______________________ Class ____________ Date ____________

Electron Configuration and Periodic Properties

Section Review **5.3**

DIRECTIONS: Write on the line at the right of each statement the letter preceding the word or expression that best completes the statement.

1. When an electron is added to a neutral atom, an amount of energy is (a) always absorbed; (b) always released; (c) either released or absorbed; (d) burned away. ______ 1
2. The energy required to remove an electron from an atom is the atom's (a) electron affinity; (b) electron energy; (c) electronegativity; (d) ionization energy. ______ 2
3. The ionization energies for removing successive electrons from sodium are 119, 1090, 1652, and 2281 kcal/mol. The great jump in ionization energy after removal of the first electron indicates that (a) sodium has 4 or 5 electrons; (b) the atomic radius increased; (c) a *d*-electron was removed; (d) the noble gas configuration has been reached. ______ 3
4. The force of attraction by Group 1 metals for their valence electrons is (a) weak; (b) zero; (c) strong; (d) greater than that for inner shell electrons. ______ 4
5. The electron-configuration notation for iron is $1s^22s^22p^63s^23p^63d^64s^2$. The third electron removed from an iron atom is one that is designated (a) $3s$; (b) $3p$; (c) $3d$; (d) $4s$. ______ 5

DIRECTIONS: Write on the line at the right of each statement the word or expression that best completes the meaning when substituted for the corresponding number.

6. When determining the "size" of an atom by measuring the distance between adjacent nuclei, the radius of an atom is __(6)__ the distance between nuclei. __________ 6
7. __(7)__ is the element in the halogen family that has the highest ionization energy. __________ 7
8. As the atomic number increases, the atomic radii of atoms tend generally to __(8)__ when moving across the periodic table. __________ 8
9. __(9)__ electrons are available to be lost, gained, or shared in the formation of chemical compounds. __________ 9
10. The valence electrons are in sublevel __(10)__ in Group 2 elements. __________ 10
11. Halogens tend to form negative ions by gaining __(11)__ electron(s). __________ 11
12. Oxygen has the electron configuration [He] $2s^22p^4$. Oxygen must gain __(12)__ electron(s) to form an ion with a charge of -2. __________ 12
13. As with main-group elements, ionization energies of *d*-block elements generally __(13)__ across a period. __________ 13
14. The first electrons to be removed in *d*-block elements during ion formation are the __(14)__ electrons. __________ 14
15. Write the electron configuration for the Na^+ ion (atomic number 11).

15

Name ______________________ Class __________ Date __________

Introduction to Chemical Bonding

Section Review 6.1

DIRECTIONS: Write on the line at the right of each statement the letter preceding the word or expression that best completes the statement.

1. In a chemical bond, the link between atoms results from the attraction between electrons and (a) Lewis structures; (b) nuclei; (c) van der Waals forces; (d) isotopes. ______ 1
2. A covalent bond consists of (a) a shared electron; (b) a shared electron pair; (c) two electrovalent ions; (d) an octet of electrons. ______ 2
3. If two covalently bonded atoms are identical, the bond is identified as (a) nonpolar covalent; (b) polar covalent; (c) nonionic; (d) coordinate covalent. ______ 3
4. A covalent bond in which there is an unequal attraction for the shared electrons is (a) nonpolar; (b) polar; (c) ionic; (d) dipolar. ______ 4
5. Atoms with a strong attraction for electrons they share with another atom exhibit (a) zero electronegativity; (b) low electronegativity; (c) high electronegativity; (d) Lewis electronegativity. ______ 5
6. Bonds with between 5% and 50% ionic character are considered to be (a) ionic; (b) pure covalent; (c) polar covalent; (d) nonpolar covalent. ______ 6
7. A nonpolar covalent bond is likely to exist between (a) a metal and a nonmetal; (b) two ions; (c) two identical atoms; (d) an atom and an ion. ______ 7
8. The greater the electronegativity difference between two bonded atoms, the greater the percentage of (a) ionic character; (b) covalent character; (c) metallic character; (d) electron sharing. ______ 8
9. In which of these compounds is the bond between the atoms NOT a nonpolar covalent bond? (a) Cl_2 (b) H_2 (c) HCl (d) O_2 ______ 9

DIRECTIONS: Complete the following statements, forming accurate sentences.

10. The electrons involved in the formation of a chemical bond are called ______________________. 10

11. A chemical bond resulting from electrostatic attraction between positive and negative ions is called a(n) ______________________. 11

12. If the electrons involved in bonding spend most of the time close to one atom rather than the other, the bond is ______________________. 12

DIRECTIONS: Questions 13 and 14 refer to the following graph.

% Ionic Character	100%	50%	5%	0%
Difference in Electronegativity	4.0	1.7	0.3	0.0

Ionic (4.0–1.7) | Polar Covalent (1.7–0.3) | Nonpolar Covalent (0.3–0.0)

13. The percentage of ionic character and the type of bond for the Li–Cl bond in LiCl (electronegativity for Li = 1.0; electronegativity for Cl = 3.0) is ______________________. 13

14. The percentage of ionic character and the type of bond for the Br–Br bond in Br_2 (electronegativity for Br = 2.8) is ______________________. 14

Name ____________________ Class __________ Date __________

Covalent Bonding and Molecular Compounds

Section Review 6.2

DIRECTIONS: Write on the line at the right of each statement the letter preceding the word or expression that best completes the statement.

1. Of the following reactions, the exothermic reaction is
(a) 1 mole $H_2 \rightarrow$ 2 moles H;
(b) 1 mole $Cl_2 \rightarrow$ 2 moles Cl;
(c) 1 mole H + 1 mole Cl $\rightarrow$ 1 mole HCl;
(d) 1 mole HCl $\rightarrow$ 1 mole H + 1 mole Cl. ______ 1

2. When an atom of one element combines chemically with an atom of another element, both atoms usually attain the stable highest-energy-level configuration of a (a) metal; (b) nonmetal; (c) noble gas; (d) metalloid. ______ 2

3. In a molecule of fluorine, the two shared electrons give each fluorine atom ______ electrons in the outer energy level. (a) 1 (b) 2 (c) 8 (d) 32 ______ 3

4. In writing a Lewis structure, each nonmetal atom except hydrogen should be surrounded by (a) 2 electrons; (b) 4 electrons; (c) 8 electrons; (d) 10 electrons. ______ 4

5. In writing a Lewis structure, the central atom is the (a) atom with the greatest mass; (b) atom with the highest atomic number; (c) atom with the fewest electrons; (d) least electronegative atom. ______ 5

6. To draw the electron-dot symbols for each atom in a Lewis structure, one must know the (a) number of valence electrons in each atom; (b) atomic mass of each atom; (c) bond length of each atom; (d) electronegativity of each atom. ______ 6

7. In the Lewis structure for the ammonium ion, there are ____ valence electrons. (a) 2 (b) 4 (c) 8 (d) 12 ______ 7

8. If, after drawing a Lewis structure, too many valence electrons have been used, the molecule probably contains (a) too many atoms; (b) one or more multiple covalent bonds; (c) too many lone electron pairs; (d) an ionic bond. ______ 8

9. The substance whose Lewis structure indicates three covalent bonds is (a) H_2O; (b) CH_2Cl_2; (c) NH_3; (d) CCl_4. ______ 9

10. How many double bonds are in the Lewis structure for hydrogen fluoride, which contains one hydrogen atom and one fluorine atom? (a) none (b) one (c) two (d) three ______ 10

11. The phosphate ion, PO_4^{3-}, contains how many extra electrons in its Lewis structure? (a) 0 (b) 2 (c) 3 (d) 4 ______ 11

DIRECTIONS: Complete the following statements, forming accurate sentences.

12. According to the equation 1 mole H_2O + 222 kcal $\rightarrow$ 2 moles H + 1 mole O, the O–H bond energy is ______________________________. 12

13. In the formation of a covalent bond, as the distance between two atoms begins to decrease, the potential energy ______________________________. 13

14. When sodium (electron configuration $1s^2 2s^2 2p^6 3s^1$) combines with chlorine to form sodium chloride, the sodium attains the electron configuration ______________________________. 14

15. The bonds within polyatomic ions are predominantly ______________________________. 15

Name ______________________ Class ____________ Date ____________

Ionic Bonding and Ionic Compounds

Section Review 6.3

DIRECTIONS: Write on the line at the right of each statement the letter preceding the word or expression that best completes the statement.

1. In the formula unit for sodium chloride, NaCl stands for one (a) formula unit; (b) molecule; (c) crystal; (d) atom. _______ 1
2. The chemical formula for an ionic compound represents the (a) number of atoms in each molecule; (b) number of ions in each molecule; (c) simplest ratio of the combined ions that gives neutrality; (d) total number of ions in the crystal lattice. _______ 2
3. A formula that shows the types and numbers of atoms combined in a single molecule is called a(n) (a) molecular formula; (b) ionic formula; (c) Lewis structure; (d) covalent formula. _______ 3
4. In a crystal of an ionic compound, each cation is surrounded by a number of (a) molecules; (b) positive ions; (c) dipoles; (d) anions. _______ 4
5. In a crystal, the valence electrons of adjacent ions (a) repel each other; (b) attract each other; (c) neutralize each other; (d) have no effects on each other. _______ 5
6. Compared to the neutral atoms involved in its formation, the crystal lattice that results is (a) higher in potential energy; (b) lower in potential energy; (c) equal in potential energy; (d) unstable. _______ 6
7. The lattice energy of compound A is greater than that of compound B. What can be deduced from this fact? (a) Compound A is not an ionic compound. (b) It will be more difficult to break the bonds in compound A than in compound B. (c) Compound B is probably a gas. (d) Compound A has larger crystals than compound B. _______ 7
8. Which of the following is NOT a property of an ionic compound? (a) vaporizes readily at room temperature (b) brittle (c) hard (d) electrical conductor in the molten state _______ 8
9. Compared to ionic compounds, molecular compounds (a) have higher boiling points; (b) are brittle; (c) have lower melting points; (d) are harder. _______ 9
10. The forces of attraction between molecules in a molecular compound are (a) stronger than the forces of ionic bonding; (b) weaker than the forces of ionic bonding; (c) approximately equal to the forces of ionic bonding; (d) zero. _______ 10
11. At room temperature, most ionic compounds will be (a) solids; (b) liquids; (c) gases; (d) molten. _______ 11

DIRECTIONS: Write on the line at the right of each statement the word or expression that best completes the meaning when substituted for the corresponding number.

12. A(n) __(12)__ is a shorthand representation of the composition of a substance using atomic symbols and numerical subscripts. _______ 12
13. In the NaCl crystal, the packing of Na^+ ions and Cl^- ions is such that each ion has clustered around it __(13)__ of the oppositely charged ions. _______ 13
14. In an ionic compound, the orderly arrangement of ions in a crystal is the state of __(14)__ energy. _______ 14
15. __(15)__ energy is the type of energy released when one mole of an ionic crystalline compound is formed from gaseous ions. _______ 15

Name ______________________ Class __________ Date __________

Metallic Bonding

Section Review 6.4

DIRECTIONS: Write on the line at the right of each statement the letter preceding the word or expression that best completes the statement.

1. Compared to nonmetals, the number of valence electrons in metals tends to be (a) smaller; (b) greater; (c) about the same; (d) almost triple that of nonmetals. ______ 1
2. In metals, the valence electrons are considered to be (a) attached to particular positive ions; (b) shared by all of the atoms; (c) immobile; (d) involved in covalent bonds. ______ 2
3. In *s*-block and *d*-block metals, the number of valence electrons in the outermost *s* sublevel is usually (a) 1 or 2; (b) 2 or 3; (c) 4; (d) 8. ______ 3
4. The fact that metals are malleable and ionic crystals are brittle is best explained in terms of their (a) chemical bonds; (b) London forces; (c) heats of vaporization; (d) polarity. ______ 4
5. The property of metallic luster is most closely related to the metal's (a) electron sea; (b) covalent bonds; (c) brittle crystalline structure; (d) positive ions. ______ 5
6. As light strikes the surface of a metal, the electrons in the electron sea (a) allow the light to pass through; (b) become attached to particular positive ions; (c) fall to lower energy levels; (d) absorb and re-emit the light. ______ 6

DIRECTIONS: Write on the line at the right of each statement the word or expression that best completes the meaning when substituted for the corresponding number.

7. A __(7)__ bond is a type of chemical bond that results from the attraction between positive ions and surrounding mobile electrons. ______ 7
8. The property called __(8)__ is the ability to be shaped or extended by physical pressure. ______ 8
9. __(9)__ is the property of being able to be drawn into a wire. ______ 9
10. Metals are referred to as __(10)__ , which means shiny. ______ 10

DIRECTIONS: Fill in the spaces below by checking either metallic solid or ionic solid, depending on which type of solid is associated with the property on the left.

	Property	Metallic Solid	Ionic Solid	
11.	ductile			11
12.	brittle			12
13.	nonconductive			13
14.	malleable			14
15.	lustrous			15

Name ______________________ Class ____________ Date ____________

The Properties of Molecular Compounds

Section Review 6.5

DIRECTIONS: Write on the line at the right of each statement the letter preceding the word or expression that best completes the statement.

1. VSEPR is based on the assumption that (a) electrons in molecules repel each other; (b) positive ions attract negative ions; (c) positive ions in a crystal repel each other; (d) sp^3 hybridization occurs in metallic bonding. ______ 1
2. The structure of the methane molecule (CH_4) is described as (a) square; (b) pyramidal; (c) pentagonal; (d) tetrahedral. ______ 2
3. A molecule of ammonia is classified in VSEPR theory as AB_3E, where the E stands for a(n) (a) crystal lattice; (b) dipole; (c) hybrid orbital; (d) unshared electron pair. ______ 3
4. In molecules, unshared electron pairs tend to (a) attract valence electrons; (b) repel electrons more strongly than bonding electron-pairs; (c) repel the nucleus; (d) attract other unshared electron pairs. ______ 4
5. In orbital notation, the hybridized orbitals responsible for the bent shape of the water molecule are identified as (a) $1s^2 2s^2$; (b) ps^1; (c) sp^3; (d) $2s^2 sp^2$. ______ 5
6. Hybridization helps to explain molecular bonding in situations in which the valence electrons in the uncombined atom (a) number more than 8; (b) are in orbitals with different shapes; (c) number less than 3; (d) are not available for bond formation. ______ 6
7. Dipole-dipole forces are considered the most important forces in polar substances because the London dispersion forces (a) exist only in nonpolar substances; (b) are usually much weaker than the dipole-dipole forces; (c) are too unpredictable; (d) occur only in solids. ______ 7
8. The equal but opposite charges present in two regions of a polar molecule create a(n) (a) electron sea; (b) dipole; (c) crystal lattice; (d) ionic bond. ______ 8
9. Of the following molecules, all of which contain polar bonds, the only polar molecule is (a) CCl_4; (b) CO_2; (c) NH_3; (d) CH_4. ______ 9
10. A polar molecule is one in which (a) ions exist; (b) a region of positive charge and a region of negative charge exist; (c) only London forces exist; (d) no bonds are present. ______ 10

DIRECTIONS: Complete the following statements, forming accurate sentences.

11. According to VSEPR theory, the shape of an AB_2 molecule is ______________________. 11
12. According to VSEPR theory, the shape of an AB_3 molecule is ______________________. 12
13. The only intermolecular forces acting among atoms of noble gases and among nonpolar molecules are ______________________. 13

DIRECTIONS: Write the answers to the following on the lines provided.

14. What can VSEPR theory be used to predict? ______________________ 14
15. Write the structural formula for water, and describe its shape. ______________________ 15

Name ______________________________ Class ____________ Date ____________

Chemical Equations

Section Review 8.1

DIRECTIONS: Write on the line at the right of each statement the letter preceding the word or expression that best completes the statement.

1. The actual knowledge about what products are produced in a chemical reaction is obtained by (a) inspecting the chemical equation; (b) balancing the chemical equation; (c) laboratory analysis; (d) writing a word equation. ______ 1
2. Once the correct formula for a reactant in an equation has been written, the (a) subscripts are adjusted to balance the equation; (b) formula should not be changed; (c) same formula must appear as a product; (d) symbols used in the formula must not be used on the product side of the equation. ______ 2
3. In writing an equation in which hydrogen gas appears as a product, the correct representation of hydrogen gas is (a) H; (b) 2H; (c) H_2; (d) OH. ______ 3
4. In a chemical equation, the symbol "$\rightarrow$" is read (a) aqueous; (b) goes; (c) yields; (d) points. ______ 4
5. In a chemical equation, the upward pointing arrow is used for a(n) (a) heated reactant; (b) gaseous reactant; (c) catalyst; (d) gaseous product. ______ 5
6. How would oxygen be represented in the formula equation for the reaction of methane and oxygen to yield carbon dioxide and water? (a) oxygen (b) O (c) O_2 (d) O_3 ______ 6
7. What is the ratio of hydrogen to chlorine in the equation $H_2(g) + Cl_2(g) \rightarrow 2HCl(g)$? (a) 1:2 (b) 2:1 (c) 1:1 (d) 4:2 ______ 7
8. Balanced formula equations express all of the following EXCEPT (a) experimentally established facts; (b) mechanisms by which reactants become restructured into products; (c) identities of reactants and products in a chemical system; (d) relative quantities of reactants and products in the system. ______ 8
9. A reversible reaction is indicated by (a) a capital R; (b) a lower-case r; (c) an up and down arrow; (d) two yield symbols pointing in opposite directions. ______ 9
10. When the equation $Fe_3O_4 + Al \rightarrow Al_2O_3 + Fe$ is correctly balanced, the coefficient of Fe is (a) 3; (b) 4; (c) 6; (d) 9. ______ 10

DIRECTIONS: Write the answers to the following on the lines provided. Where appropriate, make complete statements.

11. In a chemical equation, the symbol "(s)" indicates that the substance is a(n) ______________________. 11
12. Write the word equation that represents the production of water from hydrogen and oxygen. ____________

____ 12

DIRECTIONS: Write the answers to the following in the space provided.

13. Write the formula equation for the formation of carbon dioxide from carbon and oxygen. 13

14. Balance the formula equation $NH_4NO_2 \rightarrow N_2(g) + H_2O$. 14

15. Balance the formula equation $CaO + H_2O \rightarrow Ca(OH)_2$. 15

Name ______________________ Class __________ Date __________

Types of Chemical Reactions

Section Review 8.2

DIRECTIONS: Write on the line at the right of each statement the letter preceding the word or expression that best completes the statement.

1. The reaction $2Mg(s) + O_2(g) \rightarrow 2MgO(s)$ is an example of a(n) (a) synthesis reaction; (b) decomposition reaction; (c) single replacement reaction; (d) double replacement reaction. ______ 1
2. The reaction $Mg(s) + HCl(aq) \rightarrow H_2(g) + MgCl_2(aq)$ is an example of a(n) (a) combustion reaction; (b) decomposition reaction; (c) single replacement reaction; (d) double replacement reaction. ______ 2
3. The equation $2HgO(s) \rightarrow 2Hg(l) + O_2(g)$ is an example of a(n) (a) single replacement reaction; (b) synthesis reaction; (c) ionic reaction; (d) decomposition reaction. ______ 3
4. In one type of synthesis reaction, an element combines with oxygen to yield a(n) (a) acid; (b) hydroxide; (c) oxide; (d) metal. ______ 4
5. A metal carbonate, when heated, decomposes into a metal oxide and (a) carbon; (b) carbon dioxide; (c) oxygen; (d) hydrogen. ______ 5
6. In the equation $2Al(s) + 3Fe(NO_3)_2(aq) \rightarrow 3Fe(s) + 2Al(NO_3)_3(aq)$, iron has been replaced by (a) nitrate; (b) water; (c) aluminum; (d) nitrogen. ______ 6
7. The replacement of bromine by chlorine in a salt is an example of a single replacement reaction by (a) halogens; (b) sodium; (c) water; (d) electrolysis. ______ 7
8. The reaction of calcium oxide (CaO) with water will yield (a) calcium and oxygen gas; (b) calcium hydroxide; (c) calcium and a salt; (d) carbon dioxide and water. ______ 8
9. Predict the product of the following reaction. $MgO + CO_2 \rightarrow$ (a) $MgCO_3$ (b) $Mg + CO_3$ (c) $MgC + O_3$ (d) $MgCO_2 + O$ ______ 9
10. What product(s) will result from the decomposition of HgO? (a) mercury(I) oxide (b) mercury and oxygen (c) mercury hydroxide (d) only mercury ______ 10

DIRECTIONS: Write on the line at the right of each statement the word or expression that best completes the meaning when substituted for the corresponding number.

11. A reaction in which two or more substances combine to form a new substance is called a(n) __(11)__ reaction. ______ 11
12. The equation $AX \rightarrow A + B$ is the general equation for a(n) __(12)__ reaction. ______ 12
13. A reaction in which one element replaces a similar element in a compound is called a(n) __(13)__ reaction. ______ 13
14. __(14)__ is the decomposition of a substance by an electric current. ______ 14
15. Group 1 metals react with water to produce __(15)__ and metal hydroxides. ______ 15

Name ______________________ Class ____________ Date ____________

Activity Series of the Elements

Section Review 8.3

DIRECTIONS: Write on the line at the right of each statement the letter preceding the word or expression that best completes the statement.

1. The ability of an element to react is referred to as the element's (a) valence; (b) activity; (c) stability; (d) electronegativity. ______ 1
2. An element in the activity series can replace any element (a) in the periodic table; (b) below it on the list; (c) above it on the list; (d) in its group. ______ 2
3. An activity series is useful because it allows a person to predict (a) whether or not a certain chemical reaction will occur; (b) the amount of energy released by a chemical reaction; (c) the electronegativity values of elements; (d) the melting points of elements. ______ 3
4. For a single replacement reaction, an element will replace from a compound in aqueous solution those elements (a) above it in the activity series; (b) with lower atomic masses; (c) below it in the activity series; (d) with higher atomic numbers. ______ 4
5. To replace hydrogen from steam, a metal must be (a) above cobalt in the activity series; (b) below hydrogen in the activity series; (c) a halogen; (d) less reactive than lithium. ______ 5
6. In the activity series, any metal above hydrogen reacts with acids, replacing (a) hydrogen; (b) oxygen; (c) chlorine; (d) water. ______ 6
7. The synthesis of oxides by reaction with oxygen occurs for any metal in the activity series (a) below platinum; (b) above gold; (c) above silver; (d) below tin. ______ 7
8. Oxides are formed only indirectly by metals in the activity series (a) above zinc; (b) below calcium; (c) above tin; (d) below mercury. ______ 8
9. In the presence of oxygen, O_2, gold can be expected to (a) rapidly form an oxide; (b) slowly form an oxide; (c) not react; (d) rapidly form a metallic hydroxide. ______ 9
10. In the activity series, oxides of metals below copper (a) do not exist; (b) do not decompose; (c) decompose with heat alone; (d) are acidic. ______ 10
11. In the activity series, oxides of metals below chromium yield metals when (a) heated with hydrogen; (b) cooled below 0 °C; (c) combined with oxygen; (d) added to acids. ______ 11
12. Since mercury is below copper in the activity series, heating HgO will result in (a) no reaction; (b) the formation of a hydroxide; (c) the formation of Hg_2O; (d) decomposition. ______ 12
13. Predict what will happen when calcium metal is added to a solution of magnesium chloride. (a) No reaction will occur. (b) Calcium chloride will form. (c) Magnesium calcite will form. (d) Gaseous calcium will be produced. ______ 13
14. Predict what will happen when zinc is added to water. (a) No reaction will occur. (b) Steam will be produced. (c) Zinc oxide will form. (d) Hydrogen will be released. ______ 14
15. Predict what will happen when lead is added to nitric acid. (a) No reaction will occur. (b) Oxygen will be released. (c) Lead oxide will form. (d) Hydrogen will be released. ______ 15

Name ______________________ Class __________ Date __________

Introduction to Stoichiometry

Section Review 9.1

DIRECTIONS: Write on the line at the right of each statement the letter preceding the word or expression that best completes the statement.

1. The study of the mass relationships of elements in compounds is known as (a) reaction stoichiometry; (b) composition stoichiometry; (c) percent yield; (d) Avogadro's principle. ______ 1
2. The study of the mass relationships among reactants and products in a chemical reaction is known as (a) reaction stoichiometry; (b) composition stoichiometry; (c) electron configuration; (d) periodic law. ______ 2
3. Determining how much lime (CaO) could be obtained from a known mass of limestone ($CaCO_3$) would involve the branch of chemistry known as (a) reaction stoichiometry; (b) qualitative analysis; (c) physical chemistry; (d) metallurgy. ______ 3
4. Which of the following would not be studied in the branch of chemistry called stoichiometry? (a) the mole ratio of aluminum and chlorine in aluminum chloride (b) the amount of energy required to break the ionic bonds in calcium fluoride (c) the mass of carbon produced when a known mass of sucrose decomposes (d) the number of moles of hydrogen that will react completely with a known quantity of oxygen ______ 4
5. A balanced chemical equation allows one to determine the (a) mole ratio of any two substances in the reaction; (b) energy released in the reaction; (c) electron configuration of all elements in the reaction; (d) reaction mechanism involved in the reaction. ______ 5
6. The coefficients in a chemical equation represent the (a) masses, in grams, of all reactants and products; (b) relative numbers of moles of reactants and products; (c) number of atoms in each compound in a reaction; (d) number of valence electrons involved in the reaction. ______ 6
7. The relative number of moles of hydrogen and oxygen that react to form water represents a(n) (a) mole ratio; (b) reaction sequence; (c) bond energy; (d) element proportion. ______ 7
8. If one knows the mass of reactant A in a chemical reaction, one can determine the mass of product D produced by the use of the (a) mole ratio in the chemical equation; (b) group numbers of the periodic table; (c) bond energies involved in the reaction; (d) electron configurations of the elements involved. ______ 8
9. In the reaction $2Al_2O_3(l) \rightarrow 4Al(s) + 3O_2(g)$, the mole ratio of aluminum to oxygen is (a) 10:6; (b) 5:3; (c) 2:3; (d) 4:3. ______ 9
10. In the reaction $2H_2 + O_2 \rightarrow 2H_2O$, the mole ratio of oxygen to water is (a) 1:2; (b) 2:1; (c) 8:1; (d) 1:4. ______ 10
11. In the reaction $Ca + Cl_2 \rightarrow CaCl_2$, the mole ratio of chlorine to calcium chloride is (a) 2:3; (b) 2:1; (c) 1:2; (d) 1:1. ______ 11
12. In the reaction $Zn + H_2SO_4 \rightarrow ZnSO_4 + H_2$, the mole ratio of zinc to sulfuric acid is (a) 1:6; (b) 1:1; (c) 1:2; (d) 3:1. ______ 12
13. In solving mass-mass equation problems, the coefficients from the balanced equation are used to find the (a) number of atoms that are conserved; (b) given information; (c) amount of energy involved; (d) mole proportions of the chemicals involved. ______ 13
14. A reaction stoichiometry problem in which you are given the number of moles of one substance and asked to calculate the mass of another substance is what type of problem? (a) mole–mass (b) mass–mole (c) mass–mass (d) mole–mole ______ 14
15. For the equation $A + B \rightarrow C + D$, if you are given the mass of B and asked to calculate the number of moles of C produced, you are solving a (a) mass–mass problem; (b) mole–mole problem; (c) mass–mole problem; (d) mole–mass problem. ______ 15

Name ______________________ Class __________ Date __________

Ideal Stoichiometric Calculations

Section Review 9.2

DIRECTIONS: Write the answer to questions 1–13 on the line to the right, and show your work in the space provided.

1. The equation for the Haber process for the production of ammonia is represented by the equation $N_2(g) + H_2(g) \rightarrow 2NH_3(g)$. The complete conversion of 9.0 moles of hydrogen to ammonia would require how many moles of nitrogen? __________ 1

2. In the equation $2KClO_3 \rightarrow 2KCl + 3O_2$, how many moles of oxygen are produced when 3.0 moles of $KClO_3$ decompose completely? __________ 2

3. For the reaction $C + 2H_2 \rightarrow CH_4$, how many moles of hydrogen are required to produce 10 moles of methane (CH_4)? __________ 3

4. For the reaction $2H_2 + O_2 \rightarrow 2H_2O$, how many moles of water can be produced from 6 moles of oxygen? __________ 4

DIRECTIONS: Questions 5–13 refer to the following table.

TABLE OF ATOMIC MASSES

Element	Symbol	Atomic Mass (*u*)
Hydrogen	H	1.00
Chlorine	Cl	35.45
Oxygen	O	16.00
Sulfur	S	32.06
Sodium	Na	23.00
Carbon	C	12.00
Mercury	Hg	200.59
Fluorine	F	19.00
Calcium	Ca	40.00
Cobalt	Co	58.93

5. For the reaction $2H_2 + O_2 \rightarrow 2H_2O$, approximately how many grams of water are produced from 6 moles of hydrogen? __________ 5

6. For the reaction $C + 2H_2 \rightarrow CH_4$, how many grams of hydrogen are required to produce 3 moles of methane (CH_4)? __________ 6

7. For the reaction $2HgO \rightarrow 2Hg + O_2$, how many grams of oxygen are produced from 10 moles of mercury(II) oxide? __________ 7

8. For the reaction $H_2 + F_2 \rightarrow 2HF$, how many grams of hydrogen fluoride are produced from 8 moles of fluorine? __________ 8

9. How many moles of O_2 will react with 10.0 g of H_2 to form water in the equation $2H_2 + O_2 \rightarrow 2H_2O$? __________ 9

10. For the reaction $CaO + SO_3 \rightarrow CaSO_4$, how many moles of calcium sulfate are produced from 40 g of sulfur trioxide? __________ 10

11. For the reaction $Co + F_2 \rightarrow CoF_2$, how many moles of fluorine are required to produce 290.8 g of cobalt fluoride? __________ 11

12. If 40.0 g of sulfur dioxide are formed in the reaction between sulfur and oxygen, what is the mass of oxygen used? __________ 12

13. In the equation $2NaCl + H_2SO_4 \rightarrow 2HCl + Na_2SO_4$, what is the mass of sodium chloride that reacts with 300.0 g of sulfuric acid? __________ 13

Name ______________________ Class __________ Date __________

Limiting Reactants and Percent Yield

Section Review 9.3

DIRECTIONS: Write on the line at the right of each statement the letter preceding the word or expression that best completes the statement.

1. The reactant that controls the amount of product formed in a chemical reaction is called the (a) excess reactant; (b) mole ratio; (c) composition reactant; (d) limiting reactant. _______ 1
2. In a chemical reaction, the limiting reactant (a) would be completely used first; (b) would not be completely used; (c) is unreactive; (d) must be in solution. _______ 2
3. In a chemical reaction, the reactant remaining after all of the limiting reactant is completely used is referred to as the (a) product; (b) excess reactant; (c) controlling reactant; (d) catalyst. _______ 3
4. In the reaction $A + B \rightarrow C + D$, if there is an insufficient quantity of B to completely react with all of A, then (a) A is the limiting reactant; (b) B is the limiting reactant; (c) there is no limiting reactant; (d) no product can be formed. _______ 4
5. To determine the limiting reactant in a chemical reaction, one must know the (a) available amount of one of the reactants; (b) amount of product formed; (c) available amounts of both reactants; (d) speed of the reaction. _______ 5
6. To determine the limiting reactant in a chemical reaction involving substances A and B, one could first calculate (a) the mass of 100 moles of A and B; (b) the masses of all products; (c) bond energy of A and B; (d) the amount of moles of B required to react completely with A. _______ 6
7. The maximum amount of a product that can be produced from a given amount of reactant is called the (a) percent yield; (b) mole ratio; (c) theoretical yield; (d) actual yield. _______ 7
8. In most chemical reactions the amount of product obtained is (a) equal to the theoretical yield; (b) less than the theoretical yield; (c) more than the theoretical yield; (d) more than the percent yield. _______ 8
9. A chemist interested in the efficiency of a chemical reaction would need to calculate the (a) mole ratio; (b) energy released; (c) percent yield; (d) rate of reaction. _______ 9

DIRECTIONS: Write the answer to questions 10–13 on the line to the right, and show your work in the space provided.

10. In the reaction $2H_2 + O_2 \rightarrow 2H_2O$, how many moles of water will be produced if 6 moles of hydrogen and 2 moles of oxygen are available to react? __________ 10

11. In the reaction $Mg + 2HCl \rightarrow H_2 + MgCl_2$, how many moles of magnesium chloride can be produced from 6 moles of magnesium and 8 moles of hydrochloric acid? __________ 11

DIRECTIONS: Questions 12 and 13 refer to the following table.

TABLE OF ATOMIC MASSES

Element	Symbol	Atomic Mass (u)
Hydrogen	H	1.00
Oxygen	O	16.00
Carbon	C	12.00

12. For the reaction $2H_2 + O_2 \rightarrow 2H_2O$, calculate the percent yield if 860 g of water are produced when 100 g of hydrogen react with an excess of oxygen. __________ 12

13. For the reaction $C + 2H_2 \rightarrow CH_4$, calculate the percent yield if 98 g of methane are produced when 100 g of carbon react with an excess of hydrogen. __________ 13

Name ______________________ Class ____________ Date ____________

Section Review **10.1**

Oxygen and Ozone

DIRECTIONS: Write on the line at the right of each statement the letter preceding the word or expression that best completes the statement.

1. An example of free oxygen is the oxygen (a) dissolved in water; (b) in the mineral limestone; (c) in the water molecule; (d) in the mineral sand. ______ 1
2. The ozone layer in the atmosphere helps to prevent skin cancer by blocking out much of the sun's (a) heat; (b) ultraviolet radiation; (c) infrared radiation; (d) electromagnetic radiation. ______ 2
3. The basic life process of aerobic respiration requires (a) ozone; (b) nitrogen; (c) oxygen; (d) ammonia. ______ 3
4. A molecule composed of three oxygen atoms is (a) oxygen; (b) water; (c) ozone; (d) uncombined oxygen. ______ 4
5. Oxygen and ozone are both (a) triatomic; (b) liquids at room temperature; (c) gases; (d) paramagnetic. ______ 5
6. Two unpaired electrons exist in a molecule of (a) nitrogen; (b) oxygen; (c) ozone; (d) ammonia. ______ 6
7. Experience with jet airplanes has shown that ozone in the cabin atmosphere is a(n) (a) physical irritant; (b) oxygen filter; (c) stimulant; (d) air purifier. ______ 7
8. The production of ozone from oxygen is (a) exothermic; (b) isothermic; (c) endothermic; (d) nonthermic. ______ 8
9. Oxygen of the highest purity is produced industrially (a) from liquid air; (b) from hydrogen peroxide; (c) by electrolysis of water; (d) by decomposing manganese dioxide. ______ 9
10. A method of generating oxygen is by the thermal decomposition of (a) $KClO_3$; (b) H_2O_2; (c) H_2O; (d) Na_2O_2. ______ 10
11. The most important use of oxygen is (a) for rocket propulsion; (b) for sewage treatment; (c) in producing chemical compounds such as synthetic gasoline and ammonia; (d) in the iron and steel industry. ______ 11

DIRECTIONS: Write on the line at the right of each statement the word or expression that best completes the meaning when substituted for the corresponding number.

12. The chemical formula for free oxygen is (12) . ______ 12
13. The chemical formula is (13) for ozone. ______ 13
14. In oxides, the electronegativity difference between oxygen and the other elements is greatest for elements of Group (14) . ______ 14
15. Metals are commonly welded together by the use of a torch containing (15) and acetylene. ______ 15

Name ______________________ Class ____________ Date ____________

Hydrogen

Section Review **10.2**

DIRECTIONS: Write on the line at the right of each statement the letter preceding the word or expression that best completes the statement.

1. Scientists speculate that eventually burning gas to produce heat will (a) be a source of energy, at the expense of increased pollution; (b) not be a source of energy; (c) be illegal; (d) provide an efficient, nonpolluting source of energy. ______ 1
2. Some scientists hypothesize that hydrogen nuclei were formed from (a) subatomic particles after the "big bang" occurred; (b) reactions in the cores of stars; (c) inside the core of the earth; (d) the middle layer of Jupiter and Saturn. ______ 2
3. The density of H_2 gas is so low that if it comes into existence on earth, it (a) rises up and escapes into space; (b) rises into the clouds; (c) immediately combines with oxygen to form water; (d) sinks down into the earth's core. ______ 3
4. Scientists believe that in the middle layer of Jupiter and Saturn, hydrogen exists under very high pressure in a liquid state having the properties of (a) metals; (b) nonmetals; (c) metalloids; (d) noble gases. ______ 4
5. In laboratory preparation, hydrogen is collected by water displacement because it is (a) only slightly soluble in water; (b) the gas of lowest density; (c) just as soluble in water as oxygen; (d) odorless and colorless. ______ 5
6. The correct equation for the reaction between hydrogen and oxygen to form water is (a) $H_2 + O \rightarrow H_2O$; (b) H_2O + energy $\rightarrow 2H_2O$; (c) $2H_2 + O_2 \rightarrow 2H_2O$; (d) $H_2 + O_2 \rightarrow H_2O$. ______ 6
7. Hydrogen gas is (a) the densest of all gases; (b) the least dense of all gases; (c) very soluble in water; (d) a bluish color. ______ 7
8. Hydrogen's electronegativity is most similar to that of (a) metals; (b) nonmetals; (c) noble gases; (d) metalloids. ______ 8
9. A metal that does not replace hydrogen from acids is (a) Zn; (b) Cu; (c) Fe; (d) Mg. ______ 9
10. The principal commercial method for producing hydrogen is (a) by electrolysis of water; (b) from hydrocarbons reacting with steam; (c) from water passing over hot carbon; (d) from water by replacement by a metal. ______ 10
11. When Group 1 and Group 2 metals react with water, (a) hydrogen is burned; (b) hydrogen is displaced; (c) oxygen is displaced; (d) heat is absorbed. ______ 11

DIRECTIONS: Write on the line at the right of each statement the word or expression that best completes the meaning when substituted for the corresponding number.

12. The most abundant element in the universe is __(12)__ . ____________ 12
13. Jupiter and Saturn each have a deep ocean of __(13)__ beneath their atmospheres. ____________ 13
14. The metal __(14)__ displaces hydrogen from water so vigorously that the heat of reaction ignites the hydrogen. ____________ 14
15. Hydrogen is displaced from sulfuric acid when it reacts with zinc. The equation for this reaction is $Zn + H_2SO_4 \rightarrow ZnSO_4 +$ __(15)__ . ____________ 15

Name ______________________ Class __________ Date __________

Nitrogen and Ammonia

Section Review 10.3

DIRECTIONS: Write on the line at the right of each statement the letter preceding the word or expression that best completes the statement.

1. One way that oxides of nitrogen are formed in the atmosphere is by (a) release as an end product of photosynthesis; (b) reduction of atmospheric nitrogen; (c) exhaust from combustion in automobile engines; (d) plant respiration. ______ 1
2. Nitrogen oxides present in the atmosphere can cause (a) nitrogen-fixing bacteria to become active; (b) nitrogen-fixing blue-green algae to become inactive; (c) catalysis of N_2; (d) acid rain. ______ 2
3. The planet which has a very high concentration of nitrogen in its atmosphere is (a) Saturn; (b) Mars; (c) Jupiter; (d) Earth. ______ 3
4. The high degree of solubility of ammonia in water is evidence of the (a) polar nature of the ammonia molecule; (b) nonpolar nature of the ammonia molecule; (c) acidic nature of the ammonia molecule; (d) large part of the ammonia that ionizes. ______ 4
5. Nitrogen in its diatomic form is (a) a liquid; (b) nonreactive; (c) highly reactive; (d) a pyramid shape. ______ 5
6. An ammonia molecule is (a) nonpolar; (b) polar; (c) polar covalent; (d) nonreactive. ______ 6
7. When nitrogen and oxygen are passed through an electric arc, the compound formed is (a) NO; (b) NO_2; (c) N_2O_3; (d) N_2O_4. ______ 7
8. Nitrogen is used as a blanketing atmosphere in the processing of foods to prevent (a) softening; (b) spoilage; (c) combustion; (d) loss of flavoring. ______ 8
9. The correctly balanced equation for the combustion of ammonia in air is (a) $NH_3(g) + 2O_2(g) \rightarrow HNO_3 + H_2O(l)$; (b) $4NH_3(g) + 3O_2(g) \rightarrow 2N_2(g) + 6H_2O(l)$; (c) $NH_3(g) + O_2(g) \rightarrow NO(g) + H_2(g) + H^+(aq)$; (d) $2NH_3(g) + 2O_2(g) \rightarrow 2NO(g) + 3H_2O(l)$. ______ 9
10. The most important artificial method of nitrogen-fixation involves making the compound (a) NH_3; (b) NO; (c) N_2O_3; (d) N_2H_4. ______ 10
11. The improved catalyst used in the Haber process is a mixture of oxides of potassium and aluminum and porous (a) chromium; (b) vanadium; (c) copper; (d) iron. ______ 11
12. The Haber process is carried out at temperatures of (a) 400–500 °C and 1000 atmospheres pressure; (b) 400–500 °C and 600 atmospheres pressure; (c) 1000–1500 °C and 150 atmospheres pressure; (d) 1000–1500 °C and 600 atmospheres pressure. ______ 12
13. A general term for the type of plant with which nitrogen-fixing bacteria carry on a symbiotic relationship is (a) "combining" plant; (b) nodular plant; (c) legume; (d) "fixative" plant. ______ 13
14. The presence of "free nitrogen" on earth is directly related to (a) lightning; (b) bacterial action during the nitrogen cycle; (c) plant transpiration; (d) the conversion of HNO_3 to NH_3. ______ 14

15. Define the nitrogen cycle. ____

____

____ 15

Name ______________________ Class __________ Date __________

Carbon Dioxide and Carbon Monoxide

Section Review 10.4

DIRECTIONS: Write on the line at the right of each statement the letter preceding the word or expression that best completes the statement.

1. In industrial and urban areas, carbon monoxide in the air comes mainly from (a) fireplaces; (b) automobile engines; (c) forest fires; (d) decaying algae. ______ 1
2. By volume, the earth's atmosphere is composed of about (a) 0.03% CO_2; (b) 0.3% CO_2; (c) 0.003% CO_2; (d) 3.0% CO_2. ______ 2
3. Increasing amounts of carbon dioxide in the atmosphere come from (a) automobile engines; (b) burning of coal; (c) cigarette smoke; (d) photosynthesis. ______ 3
4. Carbon dioxide is NOT used commercially for (a) removing impurities from steel; (b) leavening; (c) fire extinguishers; (d) refrigeration. ______ 4
5. Carbon dioxide molecules are (a) nonpolar; (b) polar; (c) slightly negative; (d) slightly positive. ______ 5
6. Carbon and oxygen are united in a triple bond in (a) carbon dioxide; (b) carbon monoxide; (c) glucose; (d) water. ______ 6
7. The carbon-oxygen bonds in CO_2 are (a) longer than an ordinary CO bond; (b) shorter than an ordinary CO bond; (c) ordinary CO bonds; (d) triple bonds. ______ 7
8. A reagent that can be used as a test for the presence of carbon dioxide is (a) hydrochloric acid; (b) calcium hydroxide; (c) ammonia; (d) sodium carbonate. ______ 8
9. One of the simplest ways of producing carbon dioxide in the laboratory is by (a) heating a carbonate; (b) fermentation; (c) the action of an acid on a carbonate; (d) the burning of a form of combined carbon. ______ 9
10. The reaction of iron(III) oxide with CO yields (a) Fe^{++} and CO_2; (b) Fe and CO_2; (c) Fe^{+++}; (d) Fe_3O_4. ______ 10
11. Carbon monoxide is used to reduce metal oxides to their (a) metals; (b) nitrates; (c) metallic ions; (d) hydroxides. ______ 11
12. For humans, low levels of carbon monoxide (a) have little or no physiological effect; (b) increase the amount of oxygen in cells; (c) may impair vision and reflexes; (d) cause leukemia by affecting red blood cells. ______ 12
13. Carbon monoxide is poisonous because (a) it unites readily with hemoglobin; (b) it is a reducing agent; (c) it combines with oxygen to produce carbon dioxide; (d) it readily dissolves in blood. ______ 13
14. A source of CO around the home can be (a) space heaters; (b) plants; (c) animals; (d) cooking fumes. ______ 14
15. Write the simplified equation for photosynthesis.

15

Name ______________________ Class ____________ Date ____________

The Kinetic Theory of Matter

Section Review 11.1

DIRECTIONS: Write on the line at the right of each statement the letter preceding the word or expression that best completes the statement.

1. An important idea in kinetic theory is that particles of matter (a) are in constant motion; (b) have different shapes; (c) have different colors; (d) are always fluid. _______ 1
2. The kinetic theory can explain the behavior of (a) gases only; (b) solids and liquids; (c) liquids and gases; (d) solids, liquids, and gases. _______ 2
3. The kinetic theory explains the properties of solids, liquids, and gases in terms of the energy possessed by the particles and (a) gravitational forces; (b) the forces that act between the particles; (c) diffusion; (d) the mass of the particles. _______ 3
4. If two steel balls collide, they will bounce back at the same speed as the starting rate. This is an example of (a) Boyle's law; (b) the law of gravity; (c) elastic collision; (d) Boyle's law and Charles' law. _______ 4
5. According to the kinetic theory, the most significant difference between gases and liquids is (a) the shapes of the particles; (b) the mass of each particle; (c) the distance between the particles; (d) the fact that only the collisions in liquids are elastic. _______ 5
6. An ideal gas is an imaginary gas that (a) is not made of particles; (b) conforms to all of the assumptions of the kinetic theory; (c) has particles of zero mass; (d) is made of motionless particles. _______ 6
7. One difference between a real gas and an ideal gas is that in a real gas (a) all particles move in the same direction; (b) the particles all have the same kinetic energy; (c) diffusion cannot occur; (d) the particles exert attractive forces on each other. _______ 7
8. The observation that gases can take the shape of any container is an example of (a) density; (b) expansion; (c) adhesion; (d) evaporation. _______ 8
9. If a gas with an odor is released in a room, it soon can be detected across the room because (a) the gas can diffuse; (b) the gas is dense; (c) the gas is compressed; (d) the gas condenses. _______ 9
10. The density of a substance undergoes the greatest change when (a) a liquid changes to a gas; (b) a liquid changes to a solid; (c) a solid changes to a liquid; (d) a gas cools but does not change to a liquid. _______ 10
11. According to the kinetic theory, a gas expands because (a) the particles become larger in size; (b) collisions between particles become elastic; (c) particles collide more often; (d) particles move farther apart. _______ 11
12. The conditions under which real gases most resemble ideal gases are (a) low pressure and low temperature; (b) low pressure and high temperature; (c) high pressure and high temperature; (d) high pressure and low temperature. _______ 12
13. The cases in which the kinetic theory does not hold true for gases can be explained by (a) forces between molecules; (b) differences in the kinetic energies of the molecules; (c) large spaces between the molecules; (d) the small size of real molecules. _______ 13
14. A gas at low temperature does not behave like an ideal gas because (a) there is too much space between the particles; (b) the attractive forces are too weak; (c) the kinetic energy of the particles is too low; (d) the particles undergo chemical reactions. _______ 14
15. Which of the following types of gases will behave most like an ideal gas? (a) gases made of highly polar molecules; (b) gases made of monatomic, nonpolar molecules; (c) gases made of diatomic, polar molecules; (d) gases near their condensation temperatures. _______ 15

Qualitative Description of Gases

Section Review **11.2**

DIRECTIONS: Write on the line at the right of each statement the letter preceding the word or expression that best completes the statement.

1. The constant bombardment of the walls of a container by the moving molecules of a gas produces the characteristic called (a) temperature; (b) density; (c) pressure; (d) diffusion. ______ 1
2. To fully describe the condition of a gas, one must use all of the following except (a) volume; (b) pressure; (c) temperature; (d) chemical formula. ______ 2
3. As the temperature of a gas increases, the particles of the gas (a) increase in speed; (b) increase in mass; (c) lose kinetic energy; (d) collide less frequently. ______ 3
4. A 5-liter bottle of gas on a laboratory shelf (a) contains the same number of gas particles as any other 5-liter bottle of gas; (b) contains an unknown quantity of gas; (c) has more gas particles near the top than at the bottom; (d) is at absolute zero. ______ 4
5. If you know the volume, pressure, and quantity of a gas, you can calculate its (a) density; (b) formula; (c) temperature; (d) fluidity. ______ 5
6. To study the relationship between the pressure and volume of a gas, a factor that must be held constant is the (a) density; (b) fluidity; (c) temperature; (d) collision rate. ______ 6
7. If one holds a gas at a fixed pressure and quantity, one can investigate the effect of temperature on the gas's (a) volume; (b) kinetic energy; (c) mass; (d) flammability. ______ 7
8. If the quantity and volume of a gas remain unchanged, a change in pressure will cause a change in (a) density; (b) expansion; (c) weight; (d) temperature. ______ 8
9. To study how a gas's pressure varies with the quantity of the gas, the factors that must be kept constant are (a) mass and volume; (b) mass and temperature; (c) temperature and volume; (d) mass and weight. ______ 9
10. If the pressure and temperature of a gas are held constant, one can observe the effect of a changing quantity of gas on the gas's (a) kinetic energy; (b) volume; (c) elasticity; (d) fluidity. ______ 10
11. Suppose the temperature of the air in a balloon is raised. If the pressure remains constant, what quantity must change? (a) volume (b) number of molecules (c) compressibility (d) adhesion ______ 11
12. After an automobile has been driven for several miles, the air pressure inside the tires increases. The best explanation for this is that (a) some of the air has leaked out; (b) the air particles inside begin colliding with the tire; (c) the air particles inside the tire increase their speed; (d) the atmosphere compresses the tire. ______ 12
13. The gas pressure inside a container decreases when (a) the number of gas molecules is increased; (b) the number of gas molecules is decreased; (c) the temperature is increased; (d) the number of molecules is increased and the temperature is increased. ______ 13
14. If both the temperature and volume of a gas remains constant, then (a) the pressure of the gas will remain constant; (b) the pressure change cannot be predicted; (c) the pressure will increase; (d) the pressure will decrease. ______ 14
15. If the volume of a mole of gas remains constant and the pressure increases, it must be because of (a) a temperature decrease; (b) a kinetic energy decrease; (c) condensation of the gas; (d) a temperature increase. ______ 15

Name ______________________ Class __________ Date __________

Quantitative Description of Gases

Section Review 11.3

DIRECTIONS: Write on the line at the right of each statement the letter preceding the word or expression that best completes the statement.

1. Standard temperature is exactly (a) 100 °C; (b) 273 °C; (c) 0 °C; (d) 0 K. ______ 1
2. Pressure is defined as the force per unit of (a) volume; (b) surface area; (c) length; (d) depth. ______ 2
3. Pressure and volume changes at constant temperature can be calculated using the law developed by (a) Boyle; (b) Charles; (c) Kelvin; (d) Dalton. ______ 3
4. Absolute zero is equal to (a) −760 °C; (b) −273 °F; (c) 0.01 °C; (d) −273 °C. ______ 4
5. It is believed that absolute zero is the coldest possible temperature because at that temperature (a) the volume of any gas should become zero; (b) the kinetic energy of the molecules is very large; (c) all gases diffuse; (d) the mass of the molecules becomes zero. ______ 5
6. If a gas at 0 °C were warmed to 1 °C and the pressure did not change, then the volume would (a) increase by 1/273; (b) decrease by 1/273; (c) become very large; (d) be impossible to predict. ______ 6
7. Where V is the original volume, V′ the new volume, T the original Kelvin temperature, and T′ the new Kelvin temperature, Charles' law may be expressed mathematically as (a) $V' = VT'/T$; (b) $V = V'T'/T$; (c) $V' = V - T'/T$; (d) $V' = V/T' + T$. ______ 7
8. A sample of gas at a constant volume experiences a drop in pressure of 75 mm Hg. The most likely explanation is that (a) the container exploded; (b) the temperature increased; (c) the temperature decreased; (d) fewer particles are present. ______ 8
9. If V, p, and T represent, respectively, the original volume, pressure, and temperature in the correct units, and V′, p′, and T′ represent the new conditions, the gas-law formula for this situation is (a) $pV/T' = p'V'/T$; (b) $pV'/T = p'V/T'$; (c) $p'V/T = pV'/T'$; (d) $pV/T = p'V'/T'$. ______ 9
10. The idea that the total pressure of a mixture of gases is the sum of their partial pressures is credited to (a) Charles; (b) Boyle; (c) Kelvin; (d) Dalton. ______ 10
11. To correct for the partial pressure of water vapor, the vapor pressure of H_2O at the collecting temperature is (a) divided by 22.4; (b) multiplied by 22.4; (c) subtracted from the gas pressure; (d) added to the gas pressure. ______ 11

DIRECTIONS: Write the answer to questions 12–15 on the line to the right, and show your work in the space provided.

12. A 360 mL sample of hydrogen is collected when the pressure is 800 mm of mercury. What is the volume that the gas will occupy when the pressure is 720 mm of mercury? ________ 12

13. On a cold winter morning when the temperature is −13 °C, the air pressure in an automobile tire is 1.5 atm. If the volume does not change, what will be the pressure after the tire has warmed to 13 °C? ________ 13

14. A gas collected when the temperature is 11 °C and the pressure is 710 mm mercury measures 14.8 mL. Calculate the volume of the gas at 20.0 °C and 740 mm mercury. ________ 14

15. A sample of hydrogen with a volume of 800 mL exerts a pressure of 620 mm Hg at 5 °C. Approximately what volume will it occupy at standard temperature and pressure? ________ 15

Name ______________________ Class ____________ Date ____________

Volume-Mass Relationships of Gases

Section Review 12.1

DIRECTIONS: Write on the line at the right of each statement the letter preceding the word or expression that best completes the statement.

1. Gay-Lussac recognized that at constant temperature and pressure, the volumes of gaseous reactants and products (a) always equal 1 L; (b) add up to 22.4 L; (c) equal R; (d) can be expressed as ratios of small whole numbers. ______ 1
2. The law of combining volumes applies (a) to solids, liquids, and gases; (b) only to liquids; (c) only to solids and gases; (d) only to gases. ______ 2
3. In the equation $H_2(g) + Cl_2(g) \rightarrow 2HCl(g)$, one volume of hydrogen yields how many volumes of hydrogen chloride? (a) 1 (b) 2 (c) 3 (d) 4 ______ 3
4. If 800 mL of O_2 reacts with C to produce CO_2 at constant temperature and pressure according to the equation $C + O_2(g) \rightarrow CO_2(g)$, the volume of CO_2 produced will be (a) undetermined; (b) 800 mL; (c) 400 mL; (d) 1200 mL. ______ 4
5. The volume of a gas is directly proportional to the number of moles if (a) only pressure is constant; (b) only temperature is constant; (c) either pressure or temperature is constant; (d) both pressure and temperature are constant. ______ 5
6. The principle that under similar pressures and temperatures, equal volumes of gases contain the same number of molecules is attributed to (a) Berthollet; (b) Proust; (c) Avogadro; (d) Dalton. ______ 6
7. If gas A has a molar mass greater than that of gas B and samples of each gas at identical temperatures and pressures contain equal numbers of molecules, then (a) the volumes of gas A and gas B are equal; (b) the volume of gas A is greater than that of gas B; (c) the volume of gas B is greater than that of gas A; (d) their volumes are proportional to their molar masses. ______ 7
8. Avogadro's principle led to the realization that the molecules of some substances (a) could not react; (b) were not composed of atoms; (c) were invisible; (d) were made of more than one atom. ______ 8
9. The molar volume of a gas is all of the following except (a) the volume occupied by one mole of a gas at STP; (b) equal for all gases under the same conditions; (c) 22.4 liters at STP; (d) dependent upon the size of the molecules. ______ 9
10. At STP, standard molar volume of a gas of known volume can be used to calculate the (a) number of moles of gas; (b) rate of diffusion; (c) gram-molecular weight; (d) gram-molecular volume. ______ 10
11. If the molecular formula of a gas is known, the molar volume is used directly in solving (a) mass-mass problems; (b) the volume of any mass of gas; (c) percentage composition problems; (d) gas volume-gas volume problems. ______ 11
12. Knowing the mass and volume of a gas at STP allows one to calculate the (a) identity of the gas; (b) molar mass of the gas; (c) condensation point of the gas; (d) rate of diffusion of the gas. ______ 12
13. The molar mass of a gas at STP is simply the density of the gas (a) multiplied by the mass of one mole; (b) divided by the mass of one mole; (c) multiplied by 22.4 L; (d) divided by 22.4 L. ______ 13

DIRECTIONS: Write the answer to questions 14 and 15 on the line to the right, and show your work in the space provided.

14. What is the mass of 1.00 L of ethane (C_2H_6, 30 g/mol) at STP? ______ 14

15. What is the density of carbon monoxide (CO, 28 g/mol) at STP? ______ 15

Name ______________________ Class __________ Date __________

Section Review 12.2

The Ideal Gas Law

DIRECTIONS: Write on the line at the right of each statement the letter preceding the word or expression that best completes the statement.

1. Common units for the gas constant R are (a) L × atm; (b) mol × K; (c) (L × atm)/(mol × K); (d) atm/K. _______ 1
2. The ideal gas law combines Boyle's law, Charles' law, and (a) Graham's law; (b) Avogadro's principle; (c) Gay-Lussac's law of combining volumes; (d) Dalton's principle. _______ 2
3. The value of R, the ideal gas constant, can be calculated from measured values of a gas's pressure, volume, temperature, and (a) molar amount; (b) chemical formula; (c) rate of diffusion; (d) density. _______ 3
4. All of the following equations are statements of the ideal gas law except (a) $P = nRTV$; (b) $(PV)/T = nR$; (c) $P/n = (RT)/V$; (d) $R = (PV)/(nT)$. _______ 4
5. To use the ideal gas law to determine the molar mass of a gas (a) the mass of a molar volume of the gas must be determined; (b) the mass of any known volume of the gas may be used; (c) a volume of less than 22.4 may not be used; (d) the volume measurement must be made at STP. _______ 5
6. If n and T are constant, the ideal gas law reduces to (a) Charles' law; (b) Boyle's law; (c) Avogadro's principle; (d) zero. _______ 6
7. If n and P are constant, the ideal gas law reduces to (a) Charles' law; (b) Boyle's law; (c) Avogadro's principle; (d) zero. _______ 7
8. If P and V are constant, the ideal gas law reduces to (a) Charles' law; (b) Boyle's law; (c) Avogadro's principle; (d) zero. _______ 8

DIRECTIONS: Write the answer to questions 9–15 on the line to the right, and show your work in the space provided.

9. Calculate the volume occupied by 12.0 g of carbon dioxide (CO_2, 44 g/mol) at 20.0 °C and 740 mm Hg. __________ 9
10. What is the mass of chlorine gas (Cl_2, 70.9 g/mol) contained in a 5.00 L flask at 27 °C and 720 mm Hg? __________ 10
11. Calculate the approximate volume of a 0.6 mole sample of gas at 15 °C and a pressure of 1.1 atm. __________ 11
12. What is the approximate pressure exerted by 1.2 moles of a gas with a temperature of 20 °C and a volume of 9.5 L? __________ 12
13. A gas sample, mass 0.467 g, is collected at 20 °C and 732.5 mm Hg. The volume is 200 mL. What is the molar mass of the gas? __________ 13
14. A gas sample, mass 0.686 g, is collected at 20 °C and 722.5 mm Hg. Its volume is 350 mL. What is the molar mass of the gas? __________ 14
15. A gas sample, mass 2.50 g, is collected at 20.0 °C and 732.5 mm Hg. Its volume is 1.28 L. What is the molar mass of the gas? __________ 15

Name ______________________ Class __________ Date __________

Stoichiometry of Gases

Section Review 12.3

DIRECTIONS: Write on the line at the right of each statement the letter preceding the word or expression that best completes the statement.

1. For reactants or products that are gases, the coefficients in the chemical equation indicate (a) the number of grams of each substance; (b) volume; (c) molar volume; (d) density. ______ 1
2. If the volumes of the gaseous reactants and products in the equation $2C_2H_6 + 7O_2$ yields $4CO_2 + 6H_2O$ are measured under ordinary conditions, the substance that could not be included in the volume relationship has the formula (a) C_2H_6; (b) O_2; (c) CO_2; (d) H_2O. ______ 2
3. The law that helps to explain the volume ratios in a chemical reaction is (a) Charles' law; (b) Graham's law; (c) Boyle's law; (d) Gay-Lussac's law. ______ 3
4. Volumes of gaseous reactants and products in a chemical reaction can be expressed as ratios of small whole numbers if (a) all reactants and products are gases; (b) standard temperature and pressure are maintained; (c) constant temperature and pressure are maintained; (d) all masses equal 1 mole. ______ 4
5. According to Avogadro's principle, in a chemical equation equal volumes of gases contain equal (a) numbers of molecules; (b) pressures; (c) temperatures; (d) masses. ______ 5
6. In the reaction $2C + O_2(g) \rightarrow 2CO(g)$ the volume ratio of O_2 to CO is (a) 1:1; (b) 2:1; (c) 1:2; (d) 2:2. ______ 6
7. In the reaction $2H_2(g) + O_2(g) \rightarrow 2H_2O(g)$, the volume ratio of H_2 to H_2O is (a) 1:1; (b) 2:3; (c) 4:3; (d) 4:6. ______ 7
8. The type of problem in which the atomic masses are not used in the calculations is (a) mass–mass; (b) mass–gas volume; (c) gas volume–mass; (d) gas volume–gas volume. ______ 8
9. Gas volume–gas calculations (a) must be preceded by STP corrections if the given gas is a reactant; (b) must be followed by STP corrections if the given gas is a product; (c) do not require STP corrections if constant temperature and pressure are used for gas measurements; (d) must be followed by STP corrections if both the given and required gases are products. ______ 9
10. The equation for the production of methane is $C + 2H_2(g) \rightarrow CH_4(g)$. How many liters of hydrogen are needed to produce 20 L of methane? (a) 2.0 L (b) 22.4 L (c) 40 L (d) 2 L ______ 10
11. The complete combustion of methane is given by the equation $CH_4(g) + 2O_2(g) \rightarrow 2H_2O(g) + CO_2(g)$. If 50 L of methane are burned, how many liters of carbon dioxide will be produced? (a) 50 L (b) 16.6 L (c) 25 L (d) 100 L ______ 11
12. When hydrogen burns, water vapor is produced. The equation is $2H_2(g) + O_2(g) \rightarrow 2H_2O(g)$. If 12 L of oxygen are consumed, what volume of water vapor is produced? (a) 2 L (b) 12 L (c) 24 L (d) 1 L ______ 12

DIRECTIONS: Questions 13–15 refer to the following equation.

$$H_2 + Cl_2 \rightarrow 2HCl$$

13. The volume ratio of H_2 to Cl_2 in the reaction is (a) 1:1; (b) 1:2; (c) 2:1; (d) 3:2. ______ 13
14. The volume ratio of H_2 to HCl in the reaction is (a) 1:1; (b) 1:2; (c) 2:1; (d) 2:2. ______ 14
15. The volume ratio of Cl_2 to HCl in the reaction is (a) 1:1; (b) 1:2; (c) 2:1; (d) 2:2. ______ 15

Name ______________________ Class __________ Date __________

Effusion and Diffusion

Section Review 12.4

DIRECTIONS: Write on the line at the right of each statement the letter preceding the word or expression that best completes the statement.

1. The process by which molecules of a gas randomly encounter and pass through a small opening in a container is called (a) diffusion; (b) kinetic energy; (c) fluidity; (d) effusion. ______ 1
2. According to Graham's law, the rates of diffusion of two gases at the same temperature and pressure are inversely proportional to (a) their volumes; (b) the square roots of their molar masses; (c) their compressibilities; (d) their rates of effusion. ______ 2
3. The average kinetic energy of the molecules of any gas is determined (a) by temperature; (b) by pressure; (c) by temperature and pressure; (d) by molar mass. ______ 3
4. According to Graham's law, two gases at the same temperature and pressure will have different rates of diffusion because they have different (a) volumes; (b) molar masses; (c) rates of effusion; (d) condensation points. ______ 4
5. In Graham's equation, the square roots of the molar masses can be substituted with the square roots of the (a) gas densities; (b) molar volumes; (c) compressibilities; (d) gas constants. ______ 5
6. Suppose that two gases with unequal molecular molar masses were injected into opposite ends of a long tube at the same time and allowed to diffuse toward the center. They should begin to mix (a) at the end that held the heavier gas; (b) closer to the end that held the heavier gas; (c) closer to the end that held the lighter gas; (d) exactly in the middle. ______ 6

DIRECTIONS: Questions 7–10 refer to the following table.

Element	Atomic Mass (approximate)
Argon	39.9
Bromine	79.9
Carbon	12.0
Chlorine	35.5
Fluorine	19.0
Helium	4.0
Hydrogen	1.0
Nitrogen	14.0
Oxygen	16.0

7. Compare the rates of diffusion of hydrogen and nitrogen at the same temperature and pressure. (a) Hydrogen diffuses 3.8 times faster than nitrogen. (b) Nitrogen diffuses 3.8 times faster than hydrogen. (c) Hydrogen diffuses 14 times faster than nitrogen. (d) Both gases diffuse at the same rate. ______ 7
8. How many times greater is the rate of diffusion of oxygen than that of chlorine at the same temperature and pressure? (a) 2.0 (b) 1.5 (c) 2.1 (d) 3.5 ______ 8
9. How many times greater is the rate of diffusion of hydrogen than that of chlorine at the same temperature and pressure? (a) 9.3 (b) 35 (c) 16 (d) 6 ______ 9
10. How many times greater is the rate of effusion of oxygen than that of carbon dioxide at the same temperature and pressure? (a) 1.2 (b) 2.1 (c) 12 (d) they are equal ______ 10

Name ______________________ Class __________ Date __________

Liquids

Section Review 13.1

DIRECTIONS: Write on the line at the right of each statement the letter preceding the word or expression that best completes the statement.

1. Compared to the particles in a gas, the particles in a liquid (a) have more energy; (b) are larger; (c) move around less; (d) are widely separated. ______ 1
2. The interactive forces between particles in a liquid involve all of the following except (a) London forces; (b) hydrogen bonding; (c) dipole-dipole attractions; (d) gravitational forces. ______ 2
3. The particles in both gases and liquids (a) consist only of atoms; (b) can change positions with other particles; (c) vibrate in fixed positions; (d) are packed closely together. ______ 3
4. The entropy of a liquid is generally (a) less than that of a gas; (b) more than that of a gas; (c) equal to that of a gas; (d) zero. ______ 4
5. The interactive forces between particles are (a) less effective in solids than in liquids; (b) more effective in gases than in solids; (c) equally effective in gases and in liquids; (d) more effective in liquids than in gases. ______ 5
6. The particles in a liquid have (a) less energy than those in a solid; (b) more energy than those in a gas; (c) less energy than those in a gas; (d) no kinetic energy. ______ 6
7. The attractive forces in a liquid are (a) strong enough to prevent the particles from changing positions; (b) too weak to hold the particles in fixed positions; (c) more effective than those in a solid; (d) too weak to limit the movements of the particles. ______ 7
8. The escape of high-energy molecules from the surface of a liquid is called (a) diffusion; (b) boiling; (c) evaporation; (d) equilibrium. ______ 8
9. Liquids take the shape of their containers because the attractive forces between particles are balanced by their (a) polar forces; (b) crystal structures; (c) kinetic energy; (d) vapor pressure. ______ 9
10. Liquids have all the following properties except (a) definite volume; (b) fluidity; (c) ability to diffuse; (d) easy compressibility. ______ 10
11. Liquids, like gases, are fluid because (a) the attractive forces hold the particles in fixed positions; (b) the particles constantly move about, changing positions; (c) the energy of the particles is too low to overcome the attractive forces; (d) the particles cannot be compressed. ______ 11
12. A substance is denser in the liquid state than in the gaseous state because (a) the particles are packed more closely together; (b) there are more particles for the same amount of mass; (c) the particles are more massive; (d) the attractive forces are weaker in the liquid state. ______ 12
13. The forces exerted by the particles in a liquid on the particles of a substance added to the liquid help explain (a) surface tension of the liquid; (b) evaporation of the liquid; (c) relatively high density of the liquid; (d) dissolving ability of the liquid. ______ 13
14. Increasing the pressure on a liquid results in (a) little decrease in volume; (b) a large increase in volume; (c) a large decrease in volume; (d) a change to the gaseous state. ______ 14
15. List some properties of a liquid. ____

____

____

____ 15

Name ______________________ Class ____________ Date ____________

Solids

Section Review 13.2

DIRECTIONS: Write on the line at the right of each statement the letter preceding the word or expression that best completes the statement.

1. The pattern of points that describes the arrangement of particles in a crystal structure is the (a) unit cell; (b) cube; (c) crystal lattice; (d) kind of symmetry. _______ 1
2. The energy of the particles in a solid is (a) higher than the energy of particles in a gas; (b) high enough to allow the particles to interchange with other particles; (c) higher than the energy of particles in a liquid; (d) lower than the energy of the particles in liquids and gases. _______ 2
3. The entropy of solids is (a) lower than the entropy of liquids and gases; (b) higher than entropy of liquids; (c) about equal to entropy of liquids; (d) zero. _______ 3
4. Which statement about the solid state is false? (a) the particles vibrate in fixed positions (b) the interactive forces between particles are weak (c) the entropy is lower than that of liquids (d) the particles are closer together than those in gases _______ 4
5. All solids have (a) definite shape; (b) definite volume; (c) very slow rates of diffusion; (d) all of these properties. _______ 5
6. Solids are not fluid because (a) of weak attractive forces; (b) gravity does not act on solids; (c) the particles are in fixed positions; (d) they are colder than liquids and gases. _______ 6
7. Compared to amorphous solids, crystalline solids (a) are more fluid; (b) have a well-defined melting point; (c) change shape more easily; (d) diffuses more readily. _______ 7
8. The difference between crystalline and amorphous solids is determined by (a) temperature changes; (b) pressure when the substances are formed; (c) amount of order in particle arrangement; (d) strength of molecular forces. _______ 8
9. A substance whose solid state consists of covalent molecular crystals is (a) salt; (b) water; (c) sodium; (d) diamond. _______ 9
10. In covalent molecular crystals the covalently-bonded molecules are attracted to one another by (a) positive and negative ions; (b) Van der Waals forces; (c) positive metal cations; (d) orbitals. _______ 10

DIRECTIONS: Write on the line at the right of each statement the word or expression that best completes the meaning when substituted for the corresponding number.

11. The particles in a __(11)__ state vibrate in fixed positions. ____________ 11
12. In general, most substances are most dense in the __(12)__ state. ____________ 12
13. A property that solids share with fluids is __(13)__ . ____________ 13
14. A(n) __(14)__ crystal has valence electrons which belong to the crystal as a whole. ____________ 14
15. List some characteristics of ionic crystals. ____

____

____

____ 15

Name ______________________________ Class ______________ Date ______________

Changes of State

Section Review 13.3

DIRECTIONS: Write on the line at the right of each statement the letter preceding the word or expression that best completes the statement.

1. When melting and freezing proceed at the same rate, the system is (a) sublimated; (b) amorphous; (c) metallic; (d) in equilibrium. ______ 1
2. Whenever a liquid changes to a vapor, it (a) absorbs heat energy from its surroundings; (b) is in equilibrium with its vapor; (c) is boiling; (d) is condensing. ______ 2
3. If the concentration of vapor above a liquid is zero, then (a) no condensation can occur; (b) the rate of evaporation is high; (c) the rate of condensation is high; (d) no further evaporation can occur. ______ 3
4. In terms of the kinetic theory, what is the effect on a liquid-vapor equilibrium system if the temperature of the liquid is raised? (a) The average kinetic energy of the molecules is decreased. (b) The rate of evaporation is decreased. (c) The concentration of vapor molecules above the liquid surface is decreased. (d) Eventually, equilibrium is reestablished but at a higher vapor pressure. ______ 4
5. Le Châtelier studied (a) dynamic equilibria; (b) static equilibria; (c) lattice structure; (d) amorphous solids. ______ 5
6. If a system at equilibrium is subjected to stress, the equilibrium will be displaced (a) in the direction of the stress; (b) and immediately return to the starting point; (c) so as to relieve the stress; (d) until the temperature is lowered. ______ 6
7. The equilibrium vapor pressure of a liquid is (a) the same for all liquids; (b) the pressure exerted by a vapor in equilibrium with its liquid at a given temperature; (c) constant for a particular liquid at all temperatures; (d) measured only at 0 °C. ______ 7
8. As the temperature of a liquid increases, its equilibrium vapor pressure (a) increases; (b) decreases; (c) remains constant; (d) decreases until boiling is reached and then increases. ______ 8
9. A volatile liquid is one which (a) has strong attractive forces between particles; (b) evaporates readily; (c) has an odor; (d) is ionic. ______ 9
10. Glycerol boils at a slightly higher temperature than does water. This indicates that gylcerol's attractive forces are (a) nonexistent; (b) weaker than water; (c) the same as those of water; (d) stronger than water. ______ 10
11. A phase diagram indicates the conditions under which (a) the various states of a substance exist; (b) amorphous solids become crystalline; (c) Le Châtelier's principle no longer applies; (d) all vapors become flammable. ______ 11
12. The triple point of a substance is the conditions at which (a) a solid vaporizes; (b) states of a substance coexist at equilibrium; (c) a substance melts, freezes and boils; (d) equal amounts of solid, liquid, and gas occur. ______ 12

DIRECTIONS: Write on the line at the right of each statement the word or expression that best completes the meaning when substituted for the corresponding number.

13. During the process of __(13)__, a substance changes from a solid to a vapor without passing through the liquid phase. __________ 13
14. During boiling, the temperature of a liquid __(14)__. __________ 14
15. The conditions that are graphed in a phase diagram are __(15)__ versus pressure. __________ 15

Name ______________________ Class __________ Date __________

Water

Section Review **13.4**

DIRECTIONS: Write on the line at the right of each statement the letter preceding the word or expression that best completes the statement.

1. The bent shape of a water molecule was indicated by studying (a) the structure of ice crystals; (b) the nuclei of the oxygen atoms; (c) the positions of unshared electrons; (d) all of these. ______ 1
2. Water molecules are polar because (a) they contain two types of atoms; (b) the electrons in the covalent bonds spend more time closer to the oxygen nucleus; (c) the hydrogen bonds are weak; (d) they have covalent bonds. ______ 2
3. Unlike most liquids, when water freezes it (a) yields oxygen; (b) decreases in volume; (c) increases in volume; (d) yields hydrogen. ______ 3
4. Water changes from liquid to solid at 0 °C when the pressure is (a) one atmosphere; (b) 706 mm of Hg; (c) −1 atmosphere pressure; (d) 1.436 mm of Hg. ______ 4
5. When water is warmed above 4 °C, it (a) contracts; (b) expands; (c) maintains a constant volume; (d) increases in weight. ______ 5

DIRECTIONS: Write on the line at the right of each statement the word or expression that best completes the meaning when substituted for the corresponding number.

6. The angle between the two H–O bonds in a water molecule is __(6)__ degrees. ______ 6
7. Water molecules freeze in a definite __(7)__ pattern. ______ 7
8. In ice, H_2O molecules are held together by __(8)__ . ______ 8
9. The freezing point of water is __(9)__ degrees Celsius at standard pressure. ______ 9
10. Water has a mass of __(10)__ gram(s) at its temperature of maximum density. ______ 10

DIRECTIONS: Write the answers to the following on the lines provided.

11. Name the three physical states of water. ______________________ 11

12. Why is water so important to living organisms? ______________________ 12

13. List some of the easily observed physical properties of pure water. ______________________ 13

Name ________________ Class ________ Date ________

Types of Mixtures

Section Review 14.1

DIRECTIONS: Write on the line at the right of each statement the letter preceding the word or expression that best completes the statement.

1. Which of the following is a pure substance? (a) water (b) milk (c) gasoline (d) concrete ______ 1
2. Which of the following is a heterogeneous mixture? (a) water (b) sugar-water solution (c) whole-wheat bread (d) sugar ______ 2
3. Which of the following is a homogeneous mixture? (a) water (b) sugar-water solution (c) whole-wheat bread (d) sugar ______ 3
4. All of the following are homogeneous mixtures except (a) sodium chloride; (b) sugar-water; (c) gasoline; (d) salt-water. ______ 4
5. All of the following are heterogeneous mixtures except (a) whole-wheat bread; (b) granite; (c) tap water; (d) an oil-water mixture. ______ 5
6. Which of the following is an electrolyte? (a) sodium chloride (b) sugar (c) water (d) glass ______ 6
7. A molecular substance whose water solution conducts electricity is (a) sodium chloride; (b) hydrogen chloride; (c) sugar; (d) iron. ______ 7
8. An example of a nonelectrolyte is (a) sodium chloride; (b) hydrogen chloride; (c) sugar; (d) potassium chloride. ______ 8
9. A solution contains solute particles whose size is (a) less than 0.1 nm; (b) 0.1 nm to 1 nm; (c) 1 nm to 1000 nm; (d) greater than 1000 nm. ______ 9
10. A suspension contains solute particles whose size is (a) less than 0.1 nm; (b) 0.1 nm to 1 nm; (c) 1 nm to 1000 nm; (d) greater than 1000 nm. ______ 10

DIRECTIONS: Write on the line at the right of each statement the word or expression that best completes the meaning when substituted for the corresponding number.

11. A(n) __(11)__ mixture has components that are not easily distinguishable. ______ 11
12. A salt-water solution is an example of a(n) __(12)__ mixture. ______ 12
13. Solutions of nonelectrolytes do not contain __(13)__ . ______ 13
14. A(n) __(14)__ is a homogeneous mixture of metals. ______ 14
15. Mixtures whose particles are smaller than those in heterogeneous mixtures and larger than those in solutions are called __(15)__ . ______ 15

Name ____________________ Class __________ Date __________

The Solution Process

Section Review 14.2

DIRECTIONS: Write on the line at the right of each statement the letter preceding the word or expression that best completes the statement.

1. Which will dissolve most rapidly? (a) sugar cubes in cold water (b) sugar cubes in hot water (c) powdered sugar in cold water (d) powdered sugar in hot water ______ 1
2. Which will dissolve most slowly? (a) large salt crystals in unstirred water (b) large salt crystals in stirred water (c) small salt crystals in unstirred water (d) small salt crystals in stirred water ______ 2
3. The solubility of a substance can be expressed as (a) grams of solute; (b) grams of solvent; (c) grams of solute per 100 g of water; (d) grams of water per 100 g solute. ______ 3
4. An example of a nonpolar solvent is (a) water; (b) carbon tetrachloride; (c) both of the above; (d) none of the above. ______ 4
5. Which of the following would be soluble in water? (a) potassium nitrate (b) silver (c) benzene (d) carbon tetrachloride ______ 5
6. Two substances that are immiscible are (a) water and ammonia; (b) water and ethanol; (c) carbon tetrachloride and benzene; (d) benzene and water. ______ 6
7. Which of the following is a solvent for both polar and nonpolar solutes? (a) water (b) carbon tetrachloride (c) ethanol (d) benzene ______ 7
8. Endothermic dissolving processes (a) absorb heat and have positive heats of solution; (b) release heat and have positive heats of solution; (c) absorb heat and have negative heats of solution; (d) release heat and have negative heats of solution. ______ 8
9. Which of the following values for heat of solution involves the greatest release of energy? (a) +12.40 (b) +1.33 (c) −0.01 (d) −3.59 ______ 9
10. Ammonia is more water-soluble than expected because it (a) is nonpolar; (b) is ionic; (c) reacts with water; (d) does not react with water. ______ 10
11. A solid whose solubility decreases with increasing temperature is (a) sodium chloride; (b) cerium sulfate; (c) potassium nitrate; (d) none of the above. ______ 11
12. The solubility of a substance is 8.00 g per 100 g of water. How many grams of the substance will dissolve in 250 g of water? (a) 3.2 g (b) 16 g (c) 20 g (d) 31 g ______ 12

DIRECTIONS: Write on the line at the right of each statement the word or expression that best completes the meaning when substituted for the corresponding number.

13. Sugar is soluble in water because sugar molecules are (13) . ______ 13
14. Carbon tetrachloride and benzene are examples of (14) substances. ______ 14
15. A solution at equilibrium is called a(n) (15) solution. ______ 15

Name ______________________ Class __________ Date __________

Concentrations of Solutions

Section Review 14.3

DIRECTIONS: Write on the line at the right of each statement the letter preceding the word or expression that best completes the statement.

1. Which of the following is expressed in terms of solution volume? (a) molality (b) molarity (c) percent concentration by mass (d) all of the above ______ 1
2. The symbol *m* stands for (a) volume; (b) molality; (c) percent concentration by mass; (d) molarity. ______ 2
3. Which concentration expression is common for solutions used for practical purposes in the home, in industry, and in medicine? (a) percent concentration by mass (b) molarity (c) molality (d) all of the above are used equally ______ 3
4. Which concentration expression is most often used in chemistry? (a) percent concentration by mass (b) molarity (c) molality (d) all of the above are used equally ______ 4
5. A solution that contains a low concentration of solute must be (a) unsaturated and dilute; (b) saturated and dilute; (c) dilute, but may be saturated or unsaturated; (d) saturated, but may be dilute or concentrated. ______ 5

DIRECTIONS: Write the answer to questions 6–12 on the line to the right, and show your work in the space provided.

6. What is the molarity of a solution that contains 125 g NaCl in 4.00 L solution? ______ 6
7. What is the molality of a solution that contains 31.0 g HCl in 5.00 g water? ______ 7
8. What is the percent concentration by mass of a solution that contains 4.2 g $NaNO_3$ in 60.0 g water? ______ 8
9. How many moles of HCl are present in 0.70 L of a 0.33-*M* solution of HCl? ______ 9
10. What mass of NaCl is contained in a water solution that has a concentration of 2.48 *m* and that contains 806 g water? ______ 10
11. What is the molarity of a solution that contains 0.202 mol KCl in 7.98 L solution? ______ 11
12. What is the molality of a solution that contains 5.10 mol KNO_3 in 4.47 kg water? ______ 12

DIRECTIONS: In the parentheses at the right of each expression in the first column, write the letter of the expression in the second column that is most closely related.

13. $\frac{\text{mol solute}}{\text{L solution}}$ (   )
14. $\frac{\text{mol solute}}{\text{kg solvent}}$ (   )
15. $\frac{\text{g solute}}{\text{g solution}} \times 100$ (   )

a. density
b. percent concentration by mass
c. molarity
d. volume
e. molality

Name ______________________ Class ____________ Date ____________

Colligative Properties of Solutions

Section Review 14.4

DIRECTIONS: Write on the line at the right of each statement the letter preceding the word or expression that best completes the statement.

1. How does the temperature range over which a solution remains liquid compare to the range over which the corresponding pure solvent remains liquid? (a) the solution's range is narrower (b) the solution's range is wider (c) the two ranges are equal (d) any of the above may be true, depending on the solvent ______ 1
2. Antifreezes (a) lower both vapor pressure and freezing point; (b) lower neither vapor pressure nor freezing point; (c) lower vapor pressure but not freezing point; (d) lower freezing point but not vapor pressure. ______ 2
3. Assume that equal masses of two nonvolatile solutes of different molar masses are dissolved in two identical water samples. How do the freezing-point depressions of the two resulting solutions compare? (a) the depression for the higher-molar-mass solute will be less (b) the depression for the higher-molar-mass solute will be greater (c) the two depressions will be equal (d) any of the above might be the case, depending on the particular solute ______ 3
4. Solute molar mass can be determined by boiling-point-elevation studies, assuming that the pure solute (a) is an electrolyte; (b) is volatile; (c) is nonvolatile; (d) is liquid at the boiling temperature of water. ______ 4

DIRECTIONS: Write the answer to questions 5–11 on the line to the right, and show your work in the space provided.

5. What is the freezing point of a solution that contains 2.00 mol solute per kilogram water? ______ 5
6. What is the molality of a water solution that freezes at $-0.633\,°C$? ______ 6
7. What is the boiling-point elevation of a solution that contains 4.15 mol solute per kilogram water? ______ 7
8. What is the freezing point of a solution that contains 1.99 g of a solute of molar mass 58.5 g/mol in 58.0 g water? ______ 8
9. What is the molar mass of a solute, if dissolving 100 g of the solute in 1000 g water produces a solution that freezes at $-212\,°C$? ______ 9
10. What is the molar mass of a solute, if dissolving 0.876 g of the solute in 15.0 g water produces a solution that freezes at $-1.47\,°C$? ______ 10
11. How many grams of a solute that has a molar mass of 164 g/mol must be added to 1.330 kg water to produce a solution that freezes at $-4.13\,°C$? ______ 11

DIRECTIONS: Complete the following statements, forming accurate sentences.

12. In calculations involving colligative properties, concentrations are expressed in terms of ______________. 12
13. The units of m are ______________. 13
14. The symbol for freezing point depression is ______________. 14
15. The ability of a substance to affect colligative properties depends upon its ______________. 15

Name ______________________ Class ____________ Date ____________

Ionic Compounds in Aqueous Solution

Section Review 15.1

DIRECTIONS: Write on the line at the right of each statement the letter preceding the word or expression that best completes the statement.

1. Faraday assumed that ions (a) do not exist; (b) are formed when electricity is passed through solutions; (c) are already present in solutions that conduct electricity; (d) are not charged particles. ______ 1
2. The number of water molecules that surround a hydrated ion depends upon (a) the ion's size; (b) the ion's charge; (c) both the ion's size and the ion's charge; (d) the ion's density. ______ 2
3. When water molecules are separated, energy (a) is always absorbed; (b) is always released; (c) may be absorbed or released; (d) is burned. ______ 3
4. The hydration of which of the following involves the largest energy change? (a) a small ion of high charge (b) a small ion of low charge (c) a large ion of high charge (d) a large ion of low charge ______ 4
5. The ions $SO_4^{2-}(aq)$ and $NH_4^+(aq)$ are produced by the dissociation of the compound (a) $NH_4SO_4(s)$; (b) $(NH_4)_2SO_4(s)$; (c) $NH_4(SO_4)_2(s)$; (d) $(NH_4)_3(SO_4)_2(s)$. ______ 5
6. Which of the following would you predict is only very slightly soluble in water? (a) NaBr (b) AgBr (c) $AgNO_3$ (d) $NaNO_3$ ______ 6
7. Which of the following solids produces a solubility equilibrium whose equation shows $Ba^{2+}(aq) + SO_4^{2-}(aq)$ on the right side? (a) $Ba_2SO_4(s)$ (b) $Ba(SO_4)_2(s)$ (c) $BaSO_4(s)$ (d) $Ba_2(SO_4)_2(s)$ ______ 7
8. The net ionic equation for the precipitation reaction between copper(II) chloride and sodium hydroxide is
   (a) $Cu^{2+}(aq) + 2OH^-(aq) \rightarrow Cu(OH)_2(s)$;
   (b) $Na^+(aq) + Cl^-(aq) \rightarrow NaCl(s)$;
   (c) $Cu^{2+}(aq) + 2OH^-(aq) + 2Cl^-(aq) \rightarrow Cu(OH)_2(s) + 2Cl^-(aq)$;
   (d) $Cu^{2+}(aq) + 2Cl^-(aq) + 2Na^+(aq) + S^{2-}(aq) \rightarrow Ag_2S(s) + 2NaNO_3(s)$. ______ 8

DIRECTIONS: Questions 9 and 10 refer to the following table.

Solubility of Salts

1. Common sodium, potassium, and ammonium compounds are soluble in water.
2. Common nitrates, acetates, and chlorates are soluble.
3. Common chlorides are soluble except silver, mercury(I), and lead. [Lead(II) chloride is soluble in hot water.]
4. Common sulfates are soluble except calcium, barium, strontium, and lead.
5. Common carbonates, phosphates, and silicates are insoluble except sodium, potassium, and ammonium.
6. Common sulfides are insoluble except calcium, barium, strontium, magnesium, sodium, potassium, and ammonium.

9. Determine which of the following combinations of solutions will produce a precipitate. (a) $AgNO_3$ and K_2SO_4 (b) Na_2S and KNO_3 (c) NaCl and NH_4NO_3 (d) K_2SO_4 and $Cu(NO_3)_2$ ______ 9
10. Determine which of the following combinations of solutions will produce a precipitate. (a) Na_2SO_4 and $AgNO_3$ (b) NH_4Cl and K_2CO_3 (c) NaCl and $Mg(NO_3)_2$ (d) KNO_3 and Na_2CO_3 ______ 10

Name ______________________ Class __________ Date __________

Molecular Electrolytes

Section Review 15.2

DIRECTIONS: Write on the line at the right of each statement the letter preceding the word or expression that best completes the statement.

1. Molecular substances contain (a) ionic bonds; (b) polar covalent bonds; (c) nonpolar covalent bonds; (d) either polar covalent or nonpolar covalent bonds. ______ 1
2. Ionization of a molecule AB can best be represented as (a) $A - B \rightarrow A^+ + :B^-$; (b) $A^+B^- \rightarrow A^+ + :B^-$; (c) $A^+ + :B^- \rightarrow A^+B^-$; (d) $A^+ + :B^- \rightarrow A - B$. ______ 2
3. The hydrogen ion (a) has a charge of 2^+; (b) is a base proton; (c) has a negative charge; (d) is a bare electron. ______ 3
4. Dissolving HCl in water produces (a) H_3O^+ and Cl^-; (b) H^+ and Cl^-; (c) HCl(aq); (d) H_3O^- and Cl^+. ______ 4
5. The hydronium ion contains (a) three equivalent covalent bonds; (b) three covalent bonds of different strengths; (c) two covalent and one ionic bond; (d) two ionic and one covalent bond. ______ 5
6. In water solutions of HCl, the free H^+ ion (a) occurs in very large quantities; (b) occurs in moderate quantities; (c) occurs in small quantities; (d) does not occur. ______ 6
7. Formation of hydronium ion in HCl solution (a) is favorable, but does not release energy; (b) is unfavorable, but releases energy; (c) is favorable and releases energy; (d) is unfavorable and does not release energy. ______ 7
8. Which of the following is NOT a hydrated proton? (a) H_3O^+ (b) $H_7O_3{}^+$ (c) $H_4O_2{}^+$ (d) $H_9O_4{}^+$ ______ 8
9. Which of the following is least extensively ionized in aqueous solution? (a) HF (b) HCl (c) HBr (d) HI ______ 9
10. Which of the following metal halides is both molecular and a strong electrolyte? (a) NaCl (b) Al_2Cl_6 (c) $MgCl_2$ (d) KBr ______ 10
11. A hydrogen fluoride solution at equilibrium contains significantly more (a) H_3O^+ than F^-; (b) F^- than H_3O^+; (c) H_3O^+ than HF; (d) HF than H_3O^+. ______ 11
12. Which of the following is a weak electrolyte? (a) hydrogen chloride (b) sugar (c) sodium chloride (d) acetic acid ______ 12

DIRECTIONS: In the parentheses at the right of each description in the first column, write the letter of the compound in the second column that is most closely related.

13. ionic compound that dissociates in water ( )
14. molecular compound that ionizes in water ( )
15. molecular compound that does not ionize in water ( )

a. HCl
b. NaCl
c. C_6H_6

Name ______________________ Class ____________ Date ____________

Properties of Electrolyte Solutions

Section Review 15.3

DIRECTIONS: Write on the line at the right of each statement the letter preceding the word or expression that best completes the statement.

1. Which of the following describes the activity of a light on a conductivity apparatus used to test a solution of a strong electrolyte? (a) shine brightly (b) shine moderately (c) shine dimly (d) not shine ______ 1
2. At ordinary temperatures, the fraction of ionized water molecules is roughly two out of (a) three; (b) a hundred; (c) a thousand; (d) a billion. ______ 2
3. Pure water reveals itself to be a conductor (a) only when very sensitive conductivity tests are carried out; (b) when tested with a light bulb type conductivity apparatus; (c) under almost all circumstances; (d) under no circumstances. ______ 3
4. Compared to a 0.01-*m* sugar solution, a 0.01-*m* KCl solution has (a) the same freezing-point depression; (b) about twice the freezing-point depression; (c) the same freezing-point elevation; (d) about half the freezing-point elevation. ______ 4
5. Compared to a 0.01-*m* sugar solution, a 0.01-*m* KCl solution has (a) the same boiling-point elevation; (b) roughly twice the boiling-point elevation; (c) the same boiling-point depression; (d) roughly three times the boiling-point depression. ______ 5
6. What is the freezing-point depression of nonvolatile aqueous solutions approximately equal to? (a) 1.86 °C/*m* × molality of electrolyte (b) 1.86 °C/*m* × total molality of solute particles (c) 1.86 °C (d) 0.52 °C/*m* × total molality of solute particles ______ 6
7. The Debye-Hückel theory accounts for (a) attraction between ions in solutions; (b) repulsion between ions in solutions; (c) attractions between ions in crystals; (d) the fact that electrolyte solutions have greater freezing-point depressions than do nonelectrolyte solutions. ______ 7
8. Activity is a measure of (a) actual concentration; (b) effective concentration; (c) solution energy; (d) solution temperature. ______ 8
9. Compared to the freezing-point depression for a solution of an electrolyte that dissociates into a 3+ and a 3− ion, the freezing-point depression for an equally concentrated solution of an electrolyte that dissociates into a 1+ and a 1− ion is likely to be (a) the same; (b) slightly less; (c) much less; (d) greater. ______ 9

DIRECTIONS: Write on the line at the right of each statement the word or expression that best completes the meaning when substituted for the corresponding number.

10. The freezing-point depression of a nonvolatile aqueous electrolyte solution approximately equals the molality of solute particles times the factor __(10)__ °C/*m*. __________ 10
11. Attraction between ions in a solution is accounted for by the __(11)__ theory. __________ 11
12. When water ionizes, the products are __(12)__ and __(12)__ . __________ 12
13. Nonvolatile solutes __(13)__ freezing point. __________ 13
14. Nonvolatile solutes __(14)__ boiling point. __________ 14

15. Write the equation for the ionization of water.

15

Name ______________________ Class __________ Date __________

Acids

Section Review 16.1

DIRECTIONS: Write on the line at the right of each statement the letter preceding the word or expression that best completes the statement.

1. Citric acid is found in significant quantities in (a) lemons; (b) vinegar; (c) sour milk; (d) apples. _______ 1
2. Aqueous solutions of acids (a) always have Faraday properties; (b) conduct electricity; (c) have very high boiling points; (d) cannot be prepared. _______ 2
3. An acid that can donate three protons per molecule is called a (a) diprotic acid; (b) strong acid; (c) monoprotic acid; (d) triprotic acid. _______ 3
4. Which of the following is a binary acid? (a) H_2SO_4 (b) $HC_2H_3O_2$ (c) HBr (d) NaH _______ 4
5. An acid that has the suffix -ic produces an anion that has the (a) suffix -ate; (b) suffix -ite; (c) prefix hydro-; (d) suffix -ous. _______ 5
6. Which of the following is the electron-dot structure for chloric acid? _______ 6

   (a) H:Ö:C̈l: with :Ö: below O

   (b) H:O:C̈l:Ö: with :Ö: below Cl

   (c) H:C̈l:Ö: with :Ö: below Cl

   (d) H:Ö:C̈l:Ö:

7. The electron dot structure H:Ö:S:Ö:H (with :Ö: above and :Ö: below S) is that of (a) sulfurous acid; (b) hydrosulfuric acid; (c) sulfuric acid; (d) hyposulfurous acid. _______ 7
8. The acid that is used mainly in the manufacture of explosives, rubber, plastics, dyes, and drugs is (a) hydrochloric acid; (b) phosphoric acid; (c) nitric acid; (d) sulfuric acid. _______ 8
9. The acid used in automobile batteries is (a) hydrochloric acid; (b) phosphoric acid; (c) nitric acid; (d) sulfuric acid. _______ 9

DIRECTIONS: In the parentheses at the right of each word or expression in the first column, write the letter of the expression in the second column that is most closely related.

10. proton donor ( ) a. Lewis base
11. proton acceptor ( ) b. Brønsted acid
12. electron-pair donor ( ) c. traditional acid
13. electron-pair acceptor ( ) d. Brønsted base
14. producer of H^+ ( ) e. Lewis acid
15. producer of OH^- ( ) f. traditional base

g. Arrhenius base

h. Faraday acid

Name ______________________ Class __________ Date __________

Bases and Acid-Base Reactions

Section Review **16.2**

DIRECTIONS: Write on the line at the right of each statement the letter preceding the word or expression that best completes the statement.

1. The antacid in medications such as Maalox and Di-gel is (a) magnesium hydroxide; (b) sodium hydroxide; (c) aluminum hydroxide; (d) ammonia. ______ 1
2. The base whose water suspension is used as a laxative and antidote for ingested acids is (a) magnesium hydroxide; (b) sodium hydroxide; (c) aluminum hydroxide; (d) ammonia. ______ 2
3. A salt contains an anion from an acid and a(n) (a) electron; (b) metallic cation; (c) nonmetallic cation; (d) nonmetallic cation from a base. ______ 3
4. A Brønsted base is a(n) (a) producer of OH^- ions; (b) proton acceptor; (c) electron-pair donor; (d) electron-pair acceptor. ______ 4
5. In all reactions in which it forms a covalent bond, ammonia is a (a) Brønsted base; (b) Lewis acid; (c) Lewis base; (d) traditional acid. ______ 5
6. In the reaction $NH_3 + H_2O \rightleftarrows NH_4^+ + OH^-$, H_2O is a (a) Brønsted acid; (b) Lewis base; (c) Brønsted base; (d) traditional acid. ______ 6
7. In the reaction $H_2SO_4 + KOH \rightarrow K_2SO_4 + 2H_2O$, H_2SO_4 acts as (a) a Brønsted acid; (b) a traditional acid; (c) both a Brønsted and a traditional acid; (d) a Lewis base. ______ 7
8. Hydroxides of Group-1 metals (a) are all strong bases; (b) are all weak bases; (c) are all acids; (d) may be either strong or weak bases. ______ 8
9. In water, hydroxides of Group-2 metals (a) are all strong bases; (b) are all weak bases; (c) are all acids; (d) may be either strong or weak bases. ______ 9
10. Hydroxides of Group-2 metals (a) are all soluble; (b) are all slightly soluble; (c) are all insoluble; (d) may be either soluble, slightly soluble, or insoluble. ______ 10

DIRECTIONS: Write on the line at the right of each statement the word or expression that best completes the meaning when substituted for the corresponding number.

11. Lye is the common name for the compound __(11)__ . ____________ 11
12. Substances known as __(12)__ generally taste bitter and feel slippery. ____________ 12
13. Traditional bases produce __(13)__ ions. ____________ 13
14. A(n) __(14)__ is an electron-pair donor. ____________ 14
15. H_2O acts as a(n) __(15)__ in the reaction $Ni^{2+} + nH_2O \rightleftarrows Ni(H_2O)n^{2+}$. ____________ 15

Name ______________________ Class ____________ Date ____________

Relative Strengths of Acids and Bases

Section Review 16.3

DIRECTIONS: Write on the line at the right of each statement the letter preceding the word or expression that best completes the statement.

1. A species that remains when an acid has lost a proton is called a (a) conjugate base; (b) conjugate acid; (c) strong base; (d) strong acid. ______ 1
2. Conjugate acids and bases figure prominently in which theory of acids and bases? (a) traditional (b) Lewis (c) Brønsted (d) none of the above ______ 2
3. In the reaction $HF + H_2O \rightleftarrows H_3O^+ + F^+$, the two bases are (a) HF and H_2O; (b) HF and H_3O^+; (c) HF and F^-; (d) H_2O and F^-. ______ 3
4. In the reaction $NH_3 + HClO_3 \rightleftarrows NH_4^+ + ClO_3^-$, the conjugate base of $HClO_3$ is (a) ClO_3^-; (b) NH_3; (c) NH_4^+; (d) not shown. ______ 4
5. In the reaction $HC_2H_3O_2 + NH_3 \rightleftarrows C_2H_3O_2^- + NH_4^+$, a conjugate acid-base pair is (a) $HC_2H_3O_2$ and NH_3; (b) $C_2H_3O_2^-$ and NH_4^+; (c) $HC_2H_3O_2$ and $C_2H_3O_2^-$; (d) none of the above. ______ 5
6. A base is weak if its tendency to (a) attract a proton is great; (b) attract a proton is slight; (c) donate a proton is great; (d) donate a proton is slight. ______ 6
7. If a substance has a great tendency to give up protons, its conjugate will have a (a) great tendency to give up protons; (b) great tendency to accept protons; (c) slight tendency to give up protons; (d) slight tendency to accept protons. ______ 7
8. If ClO_4^- in the equation $NH_3 + HClO_4 \rightarrow NH_4^+ + ClO_4^-$ is a weak base, then HCO_4 is a (a) strong acid; (b) strong base; (c) weak acid; (d) weak base. ______ 8
9. In order for a proton-transfer reaction to approach completion, the acidic and basic strengths of the reactants must be (a) much lower than those of the products; (b) much higher than those of the products; (c) slightly lower than those of the products; (d) roughly equal to those of the products. ______ 9
10. Given that $HC_2H_3O_2$ is a weak acid and H_2O is a weak base, the reaction $HC_2H_3O_2 + H_2O \rightarrow H_3O^+ + C_2H_3O_2^-$ (a) goes nearly to completion; (b) favors reactants; (c) slightly favors products; (d) does not occur. ______ 10
11. Acid-base reactions that have a great tendency to proceed in the forward direction have (a) an equal tendency to proceed in the reverse direction; (b) a slightly lower tendency to proceed in the reverse direction; (c) a greater tendency to proceed in the reverse direction; (d) almost no tendency to proceed in the reverse direction. ______ 11
12. Which of the following is amphoteric? (a) H_2SO_4 (b) SO_4^{2-} (c) H^+ (d) HSO_4^- ______ 12
13. Which of the following is amphoteric? (a) H_3PO_4 (b) H^+ (c) HPO_4^{2-} (d) PO_4^{3-} ______ 13
14. In the reaction $H_2O + HSO_4^- \rightleftarrows SO_4^{2-} + H_3O^+$, HSO_4^- acts as a(n) (a) acid; (b) base; (c) spectator series; (d) salt. ______ 14
15. Explain the difference between a conjugate base and its acid. ______________________________

____

____

____ 15

Name ______________________ Class __________ Date __________

Oxides, Hydroxides, and Acids

Section Review 16.4

DIRECTIONS: Write on the line at the right of each statement the letter preceding the word or expression that best completes the statement.

1. Which of the following is a basic anhydride? (a) SO_2 (b) NaOH (c) N_2O_5 (d) K_2O ______ 1
2. Which of the following is an acid anhydride? (a) H_3PO_4 (b) Na_2O (c) P_4O_{10} (d) HNO_3 ______ 2
3. H_2SO_4 is an example of (a) an acid anhydride; (b) a basic anhydride; (c) an amphoteric substance; (d) none of the above. ______ 3
4. In the reaction $SO_3(g) + H_2O(l) \rightarrow H^+(aq) + HSO_4^-(aq)$, the acid anhydride is (a) SO_3; (b) H_2O; (c) H^+; (d) HSO_4^-. ______ 4
5. The balanced equation for the reaction of CaO with water is $CaO + H_2O \rightarrow$ (a) H_2CaO_2; (b) $CaH_2 + O_2$; (c) $Ca(OH)_2$; (d) $CaO_2 + H_2$. ______ 5
6. Where, in the periodic table, are elements that form basic anhydrides located? (a) in the left and center (b) far to the right (c) in the center only (d) toward the right, but not far to the right ______ 6
7. Strontium (Sr) could be expected to form (a) an acidic oxide; (b) a basic oxide; (c) an amphoteric oxide; (d) no oxide. ______ 7
8. Helium (He) could be expected to form (a) an acidic oxide; (b) a basic oxide; (c) an amphoteric oxide; (d) no oxide. ______ 8
9. In the reaction $Al(OH)_3 + 3H_3O^+ \rightarrow Al^{3+} + 6H_2O$, the $Al(OH)_3$ acts as a(n) (a) base; (b) acid; (c) neutral substance; (d) hydrate. ______ 9
10. In the reaction $Al(OH)_3 + OH^- \rightarrow [Al(OH)_4]^-$, $Al(OH)_3$ acts as a(n) (a) base; (b) acid; (c) neutral substance; (d) hydrate. ______ 10

DIRECTIONS: Write on the line at the right of each statement the word or expression that best completes the meaning when substituted for the corresponding number.

11. A(n) __(11)__ is an oxide that reacts with water to form an acidic solution. ______ 11
12. Soluble oxides react with water to form __(12)__ ions. ______ 12
13. SO_2 is a(n) __(13)__ in the reaction $SO_2(g) + H_2O(l) \rightarrow H_2SO_3(aq)$. ______ 13
14. Gallium (Ga) could be expected to form a(n) __(14)__ oxide. ______ 14
15. A(n) __(15)__ is a substance in which an ionic bond connects the OH or O to the rest of the formula unit. ______ 15

Name ______________________ Class __________ Date __________

Chemical Reactions of Acids, Bases, and Oxides

Section Review 16.5

DIRECTIONS: Write on the line at the right of each statement the letter preceding the word or expression that best completes the statement.

1. Dilute nonoxidizing acids react with metals in (a) composition reactions; (b) decomposition reactions; (c) single replacement reactions; (d) double replacement reactions. ______ 1
2. Which of the following is an oxidizing acid? (a) dilute H_2SO_4 (b) HNO_3 (c) HCl (d) HBr ______ 2
3. A hydroxide reacts with a nonmetal oxide to produce (a) a salt only; (b) water only; (c) either a salt or a salt and water; (d) hydrogen carbonates only. ______ 3
4. Basic metal oxides react with acidic nonmetal oxides to produce (a) hydrogen-containing salts; (b) salts and oxygen; (c) water and salts; (d) oxygen-containing salts. ______ 4
5. Metal oxides are (a) always basic; (b) always amphoteric; (c) either basic or amphoteric; (d) neither basic nor amphoteric. ______ 5
6. Oxides that react with acids to produce salts and water are (a) always basic; (b) always amphoteric; (c) either basic or amphoteric; (d) neither basic nor amphoteric. ______ 6
7. What determines whether Na_2CO_3 or $NaHCO_3$ is produced when CO_2 and NaOH react? (a) temperature (b) pressure (c) relative quantities of reactants (d) none of the above ______ 7
8. The products of the reaction between magnesium metal and dilute sulfuric acid are (a) $MgSO_3$ and H_2; (b) $MgSO_4$, H_2S, and H_2O; (c) $MgSO_4$ and H_2; (d) $MgSO_4$ and H_2O. ______ 8
9. The products of the reaction between copper metal and dilute nitric acid are (a) $Cu(NO_3)_2$, NO, and H_2O; (b) $Cu(NO_3)_2$ and H_2; (c) $Cu(NO_3)$, NO_2, and H_2O; (d) $Cu(NO_3)_2$, NO_2, and H_2O. ______ 9
10. Reactions between a nonoxidizing acid and a metal below hydrogen in the metal activity series (a) produce salts as the only product; (b) produce salts and water as the only products; (c) produce salts and hydrogen gas as the only products; (d) do not produce salts as products. ______ 10
11. A reaction between an acid and a metal oxide produces a (a) salt as the only product; (b) salt and water as the only products; (c) salt and hydrogen gas as the only products; (d) salt, water, and a gaseous compound as products. ______ 11

DIRECTIONS: In the parentheses at the right of the reactants in the first column, write the letter of the products in the second column.

Reaction Between		Produces
12. acids and carbonates	( )	a. hydrogen gas and salts
13. acidic nonmetal oxides and basic metal oxides	( )	b. hydrogen gas and carbon dioxide
14. dilute nonoxidizing acids and metals	( )	c. carbon dioxide, water, and salts
15. traditional acids and bases	( )	d. salts and water
		e. oxygen-containing salts

Name ______________________ Class ____________ Date ____________

Concentration Units for Acids and Bases

Section Review **17.1**

DIRECTIONS: Write on the line at the right of each statement the letter preceding the word or expression that best completes the statement.

1. Which of the following is NOT a concentration unit? (a) 1 molar (b) 1 equivalent (c) 1% by mass (d) 1 molal ________ 1
2. To which of the following is 1 mol H_2SO_4 chemically equivalent in a complete neutralization reaction? (a) 1 mol HCl (b) 1 equiv HCl (c) ½ equiv HCl (d) 2 equiv HCl ________ 2
3. To which of the following is 1 mol KOH chemically equivalent in a complete neutralization reaction? (a) 1 mol $Mg(OH)_2$ (b) ½ mol $Mg(OH)_2$ (c) 2 mol $Mg(OH)_2$ (d) 2 equiv $Mg(OH)_2$ ________ 3
4. In proton-transfer reactions, the mass of one equivalent of an acid is the quantity, in grams, that (a) accepts one mole of protons; (b) accepts two moles of protons; (c) supplies one mole of protons; (d) supplies two moles of protons. ________ 4
5. A 0.5-*N* solution contains (a) 0.5 equiv/L solution; (b) 0.5 equiv/kg solvent; (c) 0.5 mol/L solution; (d) 0.5 mol/kg solvent. ________ 5
6. A 1-*M* H_2SO_4 solution to be used in a complete neutralization reaction is (a) 0.5 *N*; (b) 1 *N*; (c) 2 *N*; (d) 4 *N*. ________ 6

DIRECTIONS: Write on the line at the right of each statement the number that best completes the meaning when substituted for the corresponding number.

7. One mole of H_3PO_4 in a partial neutralization reaction in which Na_2HPO_4 is formed equals ___(7)___ equiv H_3PO_4. ________ 7
8. The normality of a 1-*M* HNO_3 solution to be used in a complete neutralization reaction equals ___(8)___ *N*. ________ 8
9. The normality of a 2-*M* H_2SO_4 solution to be used in a complete neutralization reaction equals ___(9)___ *N*. ________ 9

DIRECTIONS: Write the answer to questions 10–15 on the line to the right, and show your work in the space provided.

10. What is the mass of 1 equiv H_2SO_3 in a complete neutralization reaction? ________ 10

11. What is the mass of 4.18 equiv $Mg(OH)_2$ in a complete neutralization reaction? ________ 11

12. How many equivalents are in 17.6 g KOH in a complete neutralization reaction? ________ 12

13. What is the mass of 1 equiv HBr in a complete neutralization reaction? ________ 13

14. What is the normality of a 0.312-*M* H_3PO_4 solution that is to be used in a complete neutralization reaction? ________ 14

15. How many milliliters of concentrated hydrochloric acid must be used to make 1 L of 0.240-*N* HCl? Assume that HCl is 37.5% by mass HCl and has a density of 1.19 g/mL. ________ 15

Name ________________ Class ________ Date ________

Aqueous Solutions and the Concept of pH

Section Review **17.2**

DIRECTIONS: Write on the line at the right of each statement the letter preceding the word or expression that best completes the statement.

1. Pure water contains (a) water molecules only; (b) hydronium ions only; (c) hydroxide ions only; (d) water molecules, hydronium ions, and hydroxide ions. ______ 1
2. The natural partial breakdown of water into ions is (a) autoprotolysis; (b) synthesis; (c) composition; (d) electrolysis. ______ 2
3. Which of the following equals the concentration of H_3O^+? (a) $10^{-14} - [OH^-]$ (b) $10^{-14} \times [OH^-]$ (c) $10^{-14}/[OH^-]$ (d) $[OH^-]/10^{-14}$ ______ 3
4. If $[H_3O^+]$ of a solution is greater than $[OH^-]$, the solution (a) is always acidic; (b) is always basic; (c) is always neutral; (d) may be either acidic, basic, or neutral. ______ 4
5. To calculate the pH of a solution whose $[OH^-]$ is known, one must first calculate (a) $[H_3O^+]$; (b) $\log[OH^-]$; (c) antilog$[H_3O^+]$; (d) $[H_2O]$. ______ 5
6. A water solution that has a pH of 4 (a) is always neutral; (b) is always basic; (c) is always acidic; (d) may be either neutral, acidic, or basic. ______ 6
7. A water solution that has a pH of 10 (a) is always neutral; (b) is always basic; (c) is always acidic; (d) may be either neutral, acidic, or basic. ______ 7
8. Which of the following describes chemical solutions with pH values below 0 or above 14? (a) all chemical solutions have such values (b) they are very common (c) they are rather uncommon (d) they do not exist ______ 8

DIRECTIONS: In the parentheses at the right of each expression in the first column, write the letter of the expression in the second column that is most closely related.

9. Concentration of H_3O^+ in pure water. ( )
10. Product of concentrations of H_3O^+ and OH^- in water solutions. ( )
11. Concentration of OH^- in water solutions. ( )

a. 10^{-7}
b. 10^{-14}
c. 10^{7}
d. 55.4
e. 10^{-28}
f. $\frac{10^{-14}}{[H_3O^+]}$

DIRECTIONS: Write the answer to questions 13–15 on the line to the right, and show your work in the space provided.

12. What is the pH of a solution for which $[H_3O^+] = 1.7 \times 10^{-3}$ *M*? ________

13. What is the pH of a 0.0270-*M* KOH solution? ________ 13

14. What is the hydronium-ion concentration of a solution whose pH is 4.12? ________ 14

Name ______________________ Class __________ Date __________

Acid-Base Titrations

Section Review 17.3

DIRECTIONS: Write on the line at the right of each statement the letter preceding the word or expression that best completes the statement.

1. The color-change interval for methyl orange is pH (a) 3.2 to 4.4; (b) 5.5 to 8.0; (c) 6.0 to 7.6; (d) 8.2 to 10.6. _______ 1
2. Indicator behavior can best be understood in terms of (a) the traditional theory of acids and LeChâtelier's principle; (b) the Brønsted theory of acids only; (c) the Brønsted theory of acids and LeChâtelier's principle; (d) the Lewis theory of acids only. _______ 2
3. In an indicator whose transition range is well above 7, ionization tends to be (a) more complete, and In^- is a stronger base; (b) more complete, and In^- is a weaker base; (c) less complete, and In^- is a stronger base; (d) less complete, and In^- is a weaker base. _______ 3
4. An acid-base titration is carried out by the monitoring of (a) temperature; (b) pH; (c) pressure; (d) density. _______ 4
5. In acid-base titration, equivalent quantities of hydronium and hydroxide ions are present at (a) the beginning point; (b) the midpoint; (c) the endpoint; (d) all points. _______ 5
6. A primary standard is used to (a) establish the concentration of the "known" solution; (b) titrate the "unknown" solution; (c) calibrate the pH meter; (d) drive the titration to completion. _______ 6

DIRECTIONS: Write on the line at the right of each statement the word or expression that best completes the meaning when substituted for the corresponding number.

7. Indicators change color over a range that is called its __(7)__ . _______ 7
8. The common indicator __(8)__ is very useful in studying neutralizations that involve strong acids and weak bases. _______ 8
9. Indicators tend mostly to be in the form that has the general formula __(9)__ in highly basic solutions. _______ 9
10. Indicators tend mostly to be in the form that has the general formula __(10)__ in highly acidic solutions. _______ 10

DIRECTIONS: Write the answer to questions 11–15 on the line to the right, and show your work in the space provided.

11. What is the molarity of an HCl solution if 50.0 mL of it is neutralized in a titration by 40.0 mL of 0.400-*M* NaOH? _______ 11
12. Calculate the molarity of a $Ba(OH)_2$ solution, 1900 mL of which is titrated completely by 261 mL of 0.505-*M* HNO_3. _______ 12
13. What is the molarity of an H_3PO_4 solution, 358 mL of which is completely titrated by 876 mL of 0.0102-*M* $Ba(OH)_2$ solution? _______ 13
14. What is the normal concentration of an HNO_3 solution if 412 mL of the solution is titrated by 147 mL of 2.03-*N* NaOH? _______ 14
15. What is the normal concentration of a $Ba(OH)_2$ solution, given that 86.3 mL of the solution is titrated by 113.0 mL of 0.00715-*N* H_3PO_4? _______ 15

Name ______________________ Class __________ Date __________

Thermochemistry

Section Review **18.1**

DIRECTIONS: Write on the line at the right of each statement the letter preceding the word or expression that best completes the statement.

1. Enthalpy is the same as (a) heat of formation; (b) heat of combustion; (c) heat content; (d) change in heat content. ______ 1
2. The quantity of energy released or absorbed during any chemical change is called (a) enthalpy; (b) heat of reaction; (c) heat content; (d) free energy. ______ 2
3. Compounds that have a highly negative heat of formation (a) do not exist; (b) are highly unstable; (c) are somewhat stable; (d) are highly stable. ______ 3
4. ΔH_f of compound X equals (a) X's heat of combustion minus the sum of the heats of formation of X's products of combustion; (b) the sum of the heats of formation of X's products of combustion minus X's heat of combustion; (c) X's heat of combustion minus X's heat of formation; (d) X's heat of formation minus X's heat of combustion. ______ 4
5. Energy released when bonds are formed is called (a) enthalpy; (b) heat content; (c) bond energy; (d) kinetic energy. ______ 5
6. When bonds are formed, (a) energy is always absorbed; (b) energy is always released; (c) entropy occurs; (d) evaporation occurs. ______ 6
7. The energy change involved in a reaction is related to (a) strengths of bonds only; (b) numbers of bonds only; (c) type of bonds only; (d) both strengths and numbers of bonds. ______ 7
8. In the water-gas-formation reaction $C(s) + H_2O(g) \rightarrow CO(g) + H_2(g)$, the energies involved in bond-formation and bond-breaking are such that, overall, (a) 131.3 kJ is absorbed by the system; (b) 131.4 kJ is released by the system; (c) 393.5 kJ is released by the system; (d) the system neither absorbs nor releases energy. ______ 8

DIRECTIONS: Write the answer to questions 9–11 on the line to the right, and show your work in the space provided.

9. Assuming that the products in a reaction have a total enthalpy of 458 kJ and that the reactants have a total enthalpy of 658 kJ, what is ΔH for the reaction? ______ 9

10. Assuming that the products in a reaction have a total heat content of 0 kJ and that the reactants have a total heat content of −100 kJ, what is the value of ΔH for the reaction? ______ 10

11. What is the heat of combustion of a compound, given that its heat of formation is −412 kJ/mol, and that its products have total heats of formation of −500 kJ (already adjusted for the coefficients)? ______ 11

DIRECTIONS: Complete the following statements, forming accurate sentences.

12. A calorimeter is a device used to measure ______________________. 12
13. The study of the changes in heat energy that accompany chemical reactions is called ______________. 13
14. If, during a reaction, more energy is released in bond-formation than is absorbed in bond-breaking, the reaction must be ______________________. 14
15. For elements in standard state, the value of ΔH_f is ______________________. 15

Name ______________________ Class ____________ Date ____________

Driving Force of Reactions

Section Review 18.2

DIRECTIONS: Write on the line at the right of each statement the letter preceding the word or expression that best completes the statement.

1. At high temperature, the production of water gas is (a) exothermic and spontaneous; (b) exothermic and nonspontaneous; (c) endothermic and spontaneous; (d) endothermic and nonspontaneous. ______ 1
2. The driving force of a reaction depends mostly on a change in (a) reactant's density; (b) enthalpy; (c) product's type; (d) entropy. ______ 2
3. A reaction for which $\Delta H = -500$ kJ is (a) definitely spontaneous; (b) probably spontaneous; (c) probably nonspontaneous; (d) definitely nonspontaneous. ______ 3
4. The self-mixing of gases is due to (a) favorable enthalpy change only; (b) favorable entropy change only; (c) favorable enthalpy and entropy changes; (d) favorable air pressure. ______ 4
5. An increase in entropy (a) always causes a process to be spontaneous; (b) always causes a process to be nonspontaneous; (c) tends to cause a process to be spontaneous; (d) tends to cause a process to be nonspontaneous. ______ 5
6. The reaction that produces water gas involves (a) unfavorable enthalpy change and unfavorable entropy change; (b) unfavorable enthalpy change and favorable entropy change; (c) favorable enthalpy change and unfavorable entropy change; (d) favorable enthalpy change and favorable entropy change. ______ 6
7. The net driving force of a reaction is the change in (a) free energy; (b) entropy; (c) enthalpy; (d) temperature. ______ 7
8. Free-energy change depends upon (a) change of entropy only; (b) temperature only; (c) change of enthalpy only; (d) temperature and changes of entropy and enthalpy. ______ 8
9. In ΔG calculations, temperature is expressed in (a) degrees Celsius; (b) Kelvins; (c) degrees Fahrenheit; (d) kilojoules. ______ 9
10. Spontaneity is favored by large positive values of (a) ΔG; (b) ΔH; (c) ΔS; (d) ΔA. ______ 10
11. What is the value of ΔG at 300 K for a reaction in which $\Delta H = -150$ kJ/mol, and $\Delta S = +2.00$ kJ/mol·K? (a) -750 kJ/mol; (b) -450 kJ/mol; (c) $+750$ kJ/mol; (d) $+450$ kJ/mol. ______ 11

DIRECTIONS: In the parentheses at the right of each symbol in the first column, write the letter of the expression in the second column that is most closely related.

12. *H*	( )	a. Celsius temperature
13. *T*	( )	b. free energy
14. *S*	( )	c. entropy
15. *G*	( )	d. reaction rate
		e. Kelvin temperature
		f. enthalpy

Name ______________________ Class ____________ Date ____________

The Reaction Process

Section Review 18.3

DIRECTIONS: Write on the line at the right of each statement the letter preceding the word or expression that best completes the statement.

1. The equation $H_2(g) + I_2(g) \rightarrow 2HI(g)$ is a(n) (a) overall reaction; (b) reaction mechanism; (c) reaction pathway; (d) intermediate reaction. ______ 1
2. The formation of hydrogen iodide from its elements involves a pathway of (a) one step; (b) two or three steps; (c) four or five steps; (d) five or six steps. ______ 2
3. Raising the temperature of gas particles (a) increases both collision energy and favorability of orientation; (b) increases neither collision energy nor favorability of orientation; (c) increases collision energy but not favorability of orientation; (d) increases favorability of orientation but not collision energy. ______ 3
4. The bonding of the activated complex is characteristic of (a) reactants only; (b) products only; (c) both reactants and products; (d) solids only. ______ 4
5. In an activated complex, (a) only bond-formation is occurring; (b) only bond-breaking is occurring; (c) both bond-formation and bond-breaking are occurring; (d) a catalyst is always produced. ______ 5

DIRECTIONS: Questions 6–10 refer to the following graph.

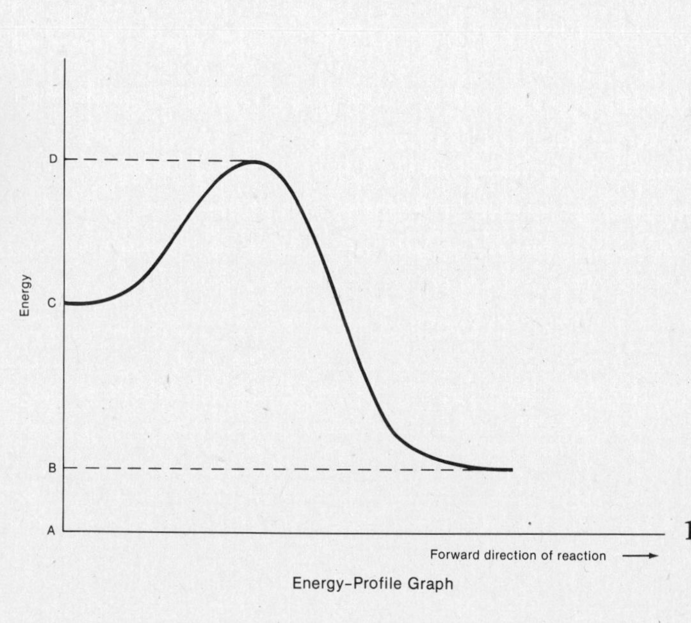

Energy–Profile Graph

6. Which energy point represents the energy of the activated complex? (a) A (b) B (c) C (d) D ______ 6
7. Which energy point represents the energy of the reactants? (a) A (b) B (c) C (d) D ______ 7
8. Which of the following is equal to the activation energy for the forward reaction? (a) C minus B (b) D minus C (c) D minus B (d) B minus C ______ 8
9. Which of the following is equal to the activation energy for the reverse reaction? (a) C minus B (b) D minus C (c) D minus B (d) B minus C ______ 9
10. Which of the following is equal to the energy of reaction for the forward reaction? (a) C minus B (b) D minus C (c) D minus B (d) B minus C ______ 10

DIRECTIONS: Complete the following statements, forming accurate sentences.

11. The overall balanced equation for the formation of hydrogen bromide from its elements is ______________________________. 11
12. Raising temperature always causes average molecular motion to ______________________________. 12
13. The minimum energy required to produce an effective collision is called a(n) ______________________________. 13
14. A short-lived structure formed during a collision is called a(n) ______________________________. 14
15. Broken lines in a representation of an activated complex represent ______________________________. 15

Name ______________________ Class ____________ Date ____________

Reaction Rate

Section Review 18.4

DIRECTIONS: Write on the line at the right of each statement the letter preceding the word or expression that best completes the statement.

1. In which type of reaction are the general conditions for reaction occurrence not always necessary? (a) synthesis (b) decomposition (c) single replacement (d) double replacement ______ 1
2. In order for most reactions to occur, it is necessary for particles (a) to collide only; (b) to be properly orientated only; (c) to have zero energy; (d) both to collide and to be properly orientated. ______ 2
3. Which factor accounts for the fact that a sample of a substance burns more rapidly in pure oxygen than in air? (a) nature of reactants (b) temperature (c) concentration (d) surface area ______ 3
4. Changing temperature affects reaction rate because such change affects (a) the energy of the activated complex; (b) the nature of the reactants; (c) the heat of reaction; (d) collision frequency and rate. ______ 4
5. Adsorption of reactants on catalyst metal surfaces changes reaction rate by affecting (a) concentration; (b) temperature; (c) the nature of the reactants; (d) volume. ______ 5
6. Catalysts generally operate by (a) increasing the temperature; (b) increasing surface area; (c) providing an alternate pathway that has a lower activation energy; (d) providing an alternate pathway that has a higher activation energy. ______ 6
7. The "poisoning" of automobile catalytic converters is caused by (a) unleaded gasoline; (b) leaded gasoline; (c) carbon monoxide; (d) nitrogen oxides. ______ 7
8. A catalyst that is heterogeneous is one that differs from the reactants in (a) mass; (b) chemical nature; (c) energy; (d) phase. ______ 8
9. If doubling the concentration of a reactant doubles rate, the concentration of the substance in the rate law has a(n) (a) exponent of 1; (b) exponent of 2; (c) exponent of 4; (d) coefficient of 2. ______ 9
10. If doubling the concentration of a reactant quadruples rate, the concentration of the substance in the rate law has a(n) (a) exponent of 1; (b) exponent of 2; (c) exponent of 4; (d) coefficient of 2. ______ 10
11. The rate law for a reaction generally depends most directly on the (a) net chemical reaction; (b) first step in the reaction pathway; (c) rate-determining step; (d) last step in the reaction pathway. ______ 11

DIRECTIONS: Complete the following statements, forming accurate sentences.

12. Reaction rate depends upon collision efficiency and collision ______________________. 12
13. Heterogeneous reactions involve reactants in different ______________________. 13
14. Decreasing temperature tends to cause the reaction rate to ______________________. 14
15. The letter R in a rate law stands for ______________________. 15

Name ______________________ Class __________ Date __________

The Nature of Chemical Equilibrium

Section Review 19.1

DIRECTIONS: Write on the line at the right of each statement the letter preceding the word or expression that best completes the statement.

1. Under suitable conditions, roughly how many of all chemical reactions are reversible? (a) none (b) less than half (c) about half (d) nearly all ______ 1
2. Which of the following equations involves reactants that cannot be reversed in a single process? (a) $2KClO_3 \rightarrow 2KCl + 3O_2$ (b) $HgO \rightarrow 2Hg + O_2$ (c) $NaCl \rightarrow Na^+ + Cl^-$ (d) $HBr + H_2O \rightarrow H_3O^+ + Br^-$ ______ 2
3. If HgO is heated in a closed container, (a) nothing occurs; (b) only decomposition occurs; (c) Hg_2O_2 is formed; (a) at first decomposition, then also recombination, occur. ______ 3
4. At equilibrium, the amount of product (a) is always equal to the amount of reactant; (b) is always greater than the amount of reactant; (c) is always less then the amount of reactant; (d) may be equal to, greater than, or less than the amount of reactant. ______ 4
5. In the expression $K = [W][X]/[Y][Z]$ the mass-action expression is (a) [Y][Z]; (b) [W][X]; (c) [W][X]/[Y][Z]; (d) $K = [W][X]/[Y][Z]$. ______ 5
6. In the expression $K = [W][X]/[Y][Z]$ the chemical equilibrium expression is (a) [Y][Z]; (b) [W][X]; (c) [W][X]/[Y][Z]; (d) $K = [W][X]/[Y][Z]$. ______ 6
7. In the expression $K = [W][X]/[Y][Z]$ the reactant concentrations are expressed as (a) [Y][Z]; (b) [W][X]; (c) [W][X]/[Y][Z]; (d) $K = [W][X]/[Y][Z]$. ______ 7
8. In the expression $K = [W][X]/[Y][Z]$ the product concentrations are expressed as (a) [Y][Z]; (b) [W][X]; (c) [W][X]/[Y][Z]; (d) $K = [W][X]/[Y][Z]$. ______ 8
9. The value of the equilibrium constant equals (a) the value of the mass-action expression; (b) the product of the concentrations of the products divided by the product of the concentrations of the reactants; (c) the product of the concentrations of the reactants divided by the product of the concentrations of the products; (d) 1, in all cases. ______ 9
10. How is the fact that a particular system very quickly reaches equilibrium reflected in the value of K? (a) K has a high value in such a case. (b) K has a low nonzero value in such a case. (c) K has a value of zero in such a case. (d) The fact is not reflected in the value of K. ______ 10
11. A value of K close to 1 indicates that it is likely that at equilibrium (a) only products will be present; (b) only reactants will be present; (c) significant quantities of both products and reactants will be present; (d) the reactions occur at a moderate rate. ______ 11
12. For cases in which the forward and reverse reactions are simple, one-step processes, how is K related to k_f and k_r from the rate laws for the forward and reverse reactions? (a) $K = k_f/k_r$ (b) $K = k_r/k_f$ (c) $K = k_f + k_r$ (d) $K = k_f - k_r$ ______ 12

DIRECTIONS: Complete the following statements, forming accurate sentences.

13. The two processes that constitute equilibrium for a saturated sugar solution are dissolving and ______________________________. 13
14. The equilibrium constant is symbolized by ______________________________. 14
15. The value of an equilibrium constant changes with ______________________________. 15

Name ______________________ Class __________ Date __________

Shifting Equilibrium

Section Review 19.2

DIRECTIONS: Write on the line at the right of each statement the letter preceding the word or expression that best completes the statement.

1. What is the effect on the quantities of substances present if the pressure on the equilibrium system $2CO(g) + O_2(g) \rightleftarrows 2CO_2(g)$ is increased? (a) the quantity of CO(g) increases (b) the quantity of $CO_2(g)$ decreases (c) the quantity of $CO_2(g)$ increases (d) there is no effect ______ 1
2. What is the effect on concentrations if the temperature of the equilibrium system $X + Y \rightleftarrows Z + 25$ kJ is decreased? (a) [X] decreases and [Z] increases (b) [X] increases and [Z] decreases (c) [X] decreases and [Z] decreases (d) there is no effect ______ 2
3. Which of the following reactions is essentially irreversible? (a) $N_2 + 3H_2 \rightarrow 2NH_3$ (b) $H_2CO_3 \rightarrow H_2O + CO_2$ (c) $2HClO_3 \rightarrow 2KCl + 3O_2$ (d) $FeS + 2HCl \rightarrow FeCl_2 + H_2S$ ______ 3
4. When dilute solutions of NaCl and KBr are mixed, (a) only NaBr precipitates; (b) only KCl precipitates; (c) both NaBr and KCl precipitate; (d) no precipitation occurs. ______ 4
5. Which of the following tends to run to completion? (a) $N_2(g) + 3H_2(g) \rightarrow 2NH_3(g)$ (b) $Na^+(aq) + Cl^-(aq) \rightarrow NaCl(s)$ (c) $2CO + O_2 \rightarrow 2CO_2$ (d) $H_2CO_3(aq) \rightarrow H_2O(l) + CO_2(g)$ ______ 5
6. Which of the following tends to run to completion? (a) $K^+(aq) + Cl^-(aq) \rightarrow KCl(s)$ (b) $2NO_2(g) \rightarrow N_2O_4(g)$ (c) $Ag^+(aq) + Cl^-(aq) \rightarrow AgCl(s)$ (d) $H_2(g) + I_2(g) \rightarrow 2HI(g)$ ______ 6
7. Which of the following tends to run to completion? (a) $N_2(g) + O_2(g) \rightarrow 2NO(g)$ (b) $H_3O^+(aq) + OH^-(aq) \rightarrow 2H_2O(l)$ (c) $Na^+(aq) + Br^-(aq) \rightarrow NaBr(s)$ (d) $H_2(g) + CO(g) \rightarrow C(s) + H_2O(g)$ ______ 7
8. The common-ion effect tends to cause (a) condensation; (b) evaporation; (c) reduced ionization; (d) increased solubility. ______ 8
9. Addition of hydrogen chloride will tend to bring about precipitation in a solution of (a) H_2SO_4; (b) KBr; (c) NaCl; (d) NaH. ______ 9
10. The addition of sodium acetate to an acetic acid ($HC_2H_3O_2$) solution (a) causes precipitation of $HC_2H_3O_2$; (b) causes precipitation of $NaC_2H_3O_2$; (c) increases ionization of $HC_2H_3O_2$; (d) decreases ionization of $HC_2H_3O_2$. ______ 10
11. Addition of hydrogen bromide to a solution of KBr will cause precipitation of (a) HBr; (b) KBr; (c) KH; (d) Br_2. ______ 11
12. Addition of sodium fluoride to a solution of the weakly ionized acid HF will (a) cause precipitation of HF; (b) cause precipitation of NaF; (c) increase ionization of HF; (d) decrease ionization of HF. ______ 12

DIRECTIONS: Complete the following statements, forming accurate sentences.

13. If the temperature of the equilibrium system $CH_3OH(g) + 101\text{ kJ} \rightleftarrows CO(g) + 2H_2(g)$ is increased, the value of K will ______________________. 13
14. An increase in the concentration of O_2 in the equilibrium system $2H_2(g) + O_2(g) \rightleftarrows 2H_2O(g)$ will cause $[H_2]$ to ______________________. 14
15. The common-ion effect tends to cause dissolved substances to ______________________. 15

Name ______________________ Class __________ Date __________

Equilibrium of Acids, Bases, and Salts

Section Review **19.3**

DIRECTIONS: Write on the line at the right of each statement the letter preceding the word or expression that best completes the statement.

1. In the equilibrium system $HC_2H_3O_2 \rightleftarrows H_3O^+ + C_2H_3O_2^-$ which reaction proceeds more rapidly? (a) the forward reaction (b) the reverse reaction (c) neither reaction occurs (d) both reactions occur at equal rates ______ 1
2. The general relationship between K and K_a is (a) $K = K_a$; (b) $K = K_a[H_3O]$; (c) $K_a = K[H_3O^+]$; (d) $K_a = [H_3O^+]/K$. ______ 2
3. A solution containing both a weak acid and a salt of the acid can react with (a) acids only; (b) bases only; (c) either acids or bases; (d) neither acids nor bases. ______ 3
4. If a base is added to a solution of weak acid and its salt, (a) more of the un-ionized weak acid is formed; (b) more of the un-ionized acid ionizes; (c) precipitation occurs; (d) hydronium-ion concentration decreases. ______ 4
5. If a base is added to a solution of a weak base and its salt, (a) hydronium-ion concentration increases; (b) more of the weak base ionizes; (c) more hydroxide ion is formed; (d) more water and unionized base are formed. ______ 5
6. An example of a salt that produces a neutral solution is (a) NaCl; (b) $NaC_2H_3O_2$; (c) NH_4Cl; (d) H_2O. ______ 6
7. The equilibrium constant for a reaction between water and dissolved ions is called the (a) ion-product constant; (b) hydrolysis constant; (c) solubility-product constant; (d) acid-ionization constant. ______ 7
8. The relationship between the ion-product constant for water, the acid-ionization constant, and the hydrolysis constant is (a) $K_h = K_w \times K_a$; (b) $K_h = K_w - K_a$; (c) $K_h = K_w/K_a$; (d) $K_h = K_a/K_w$. ______ 8
9. The dissolving of sodium carbonate in water produces a solution that is (a) always acidic; (b) always basic; (c) always neutral; (d) either acidic, basic, or neutral. ______ 9
10. The hydrolysis of NH_4Cl involves water and (a) NH_4^+ ions only; (b) Cl^- ions only; (c) both NH_4^+ and Cl^- ions; (d) neither NH_4^+ nor Cl^- ions. ______ 10
11. Dissolved hydrates of metallic cations such as $Al(H_2O)_6^{3+}$ and $Cu(H_2O)_4^{2+}$ (a) do not hydrolyze; (b) hydrolyze to produce acidic solutions only; (c) hydrolyze to produce basic solutions only; (d) ionize to produce acidic solutions only. ______ 11

DIRECTIONS: In the parentheses at the right of each expression in the first column, write the letter of the number in the second column that is most closely related.

12. concentration, in moles per liter, of H_3O^+ in pure water ( )
13. concentration, in moles per liter, of H_2O in pure water ( )
14. pH of pure water ( )
15. value of the ion-product constant for water ( )

a. 0
b. 10^{-14}
c. 10^{-7}
d. 1
e. 7
f. 14
g. 55.4

Name ______________________ Class __________ Date __________

Solubility Equilibrium

Section Review 19.4

DIRECTIONS: Write on the line at the right of each statement the letter preceding the word or expression that best completes the statement.

1. Substances whose solubilities range between 0.1 g and 1 g per 100 g water are said to be (a) soluble; (b) slightly soluble; (c) sparingly soluble; (d) insoluble. ______ 1
2. The equilibrium equation for the dissolving of silver chloride is (a) $2AgCl(aq) \rightleftarrows 2Ag(s) + Cl_2(aq)$; (b) $2AgCl(s) \rightleftarrows 2Ag(aq) + Cl_2(aq)$; (c) $AgCl(aq) \rightleftarrows Ag^+(s) + Cl^-(aq)$; (d) $AgCl(s) \rightleftarrows Ag^+(aq) + Cl^-(aq)$. ______ 2
3. Given the K_{sp} value of a compound, what additional information is needed to calculate solubility in moles per liter? (a) molar mass (b) density of the solid (c) solubility in grams (d) no other information ______ 3
4. Solubility-product principles can be applied very successfully to (a) only slightly soluble substances; (b) only insoluble substances; (c) only very soluble substances; (d) all substances. ______ 4

DIRECTIONS: Write the answer to questions 5–11 on the line to the right, and show your work in the space provided.

5. What is the solubility-product constant of magnesium hydroxide ($Mg(OH)_2$), given that the solubility of this compound is 9.0×10^{-4} g/100 g water? ______ 5
6. What is the solubility, in mol/L, of silver iodide (AgI), given that its K_{sp} value is 8.3×10^{-17}? ______ 6
7. What is the solubility, in mol/L, of copper(I) chloride, given that its K_{sp} value is 1.2×10^{-6}? ______ 7
8. What is the solubility, in mol/L, of copper(II) sulfide (CuS), given that its K_{sp} value is 6.3×10^{-36}? ______ 8
9. What is the solubility, in mol/L, of silver sulfide (Ag_2S), given that its K_{sp} value is 1.6×10^{-49}? ______ 9
10. Calculate the ion product for the mixing of 100.0 mL of 0.0030-M $CaCl_2$ with 100 mL of 0.0020-M Na_2CO_3, given that K_{sp} for $CaCO_3$ is 1.4×10^{-8}. Will precipitation occur? ______ 10
11. Calculate the ion product for the mixing of 300 mL of 0.00030-M $Sr(NO_3)_2$ with 200 mL of 0.000025-M K_2SO_4, given that K_{sp} for $SrSO_4$ is 3.2×10^{-7}. Will precipitation occur? ______ 11

DIRECTIONS: Write on the line at the right of each statement the word or expression that best completes the meaning when substituted for the corresponding number.

12. The symbol __(12)__ stands for the solubility-product constant. ______ 12
13. The mass-action expression __(13)__ equals K_{sp} for the dissolving of silver bromide (AgBr). ______ 13
14. If the ion product for a given pair of ions whose solutions have just been mixed is greater than the value of the solubility-product constant, __(14)__ occurs. ______ 14
15. Precipitation from a mixture of two solutions, if it occurs, continues until __(15)__ is established. ______ 15

Name ______________________ Class ____________ Date ____________

The Nature of Oxidation and Reduction

Section Review 20.1

DIRECTIONS: Write on the line at the right of each statement the letter preceding the word or expression that best completes the statement.

1. Numbers assigned to atoms and ions as a way of keeping track of electrons are called (a) charges; (b) coefficients; (c) electrode potentials; (d) oxidation numbers. ______ 1
2. In the reaction $O_2 + 4e^- \rightarrow 2O_2^-$, O_2 is being (a) oxidized; (b) reduced; (c) electrolyzed; (d) auto-oxidized. ______ 2
3. In the reaction $Fe^{3+} \rightarrow Fe + 3e^-$, Fe^{3+} is being (a) oxidized; (b) reduced; (c) electrolyzed; (d) auto-oxidized. ______ 3
4. In the reaction $Na^+ + Br^- \rightarrow NaBr$, which species is being reduced? (a) Na^+ only (b) Br^- only (c) both Na^+ and Br^- (d) neither Na^+ nor Br^- ______ 4
5. The oxidation number of combined hydrogen is usually equal to (a) -2; (b) -1; (c) 0; (d) $+1$. ______ 5
6. The algebraic sum of the oxidation numbers of the atoms in a compound (a) is always zero; (b) is always $+1$; (c) is always -1; (d) can be any value. ______ 6
7. What are the oxidation numbers in the compound KCl? (a) $K = 0$, $Cl = 0$ (b) $K = -1$, $Cl = +1$ (c) $K = +1$, $Cl = -1$ (d) $K = +2$, $Cl = -2$ ______ 7
8. Oxidation and reduction (a) always occur simultaneously; (b) always occur at different times; (c) do not occur in the same reaction; (d) always occur oxidation first, then reduction. ______ 8
9. Any chemical process in which elements undergo a change in oxidation number is called a(n) (a) synthesis; (b) decomposition; (c) electrolysis; (d) oxidation-reduction reaction. ______ 9
10. Which of the following products could be made from SO_3 through an oxidation-reduction reaction? (a) H_2SO_4 (b) H_2SO_3 (c) SF_6 (d) SO ______ 10
11. In a redox reaction in which CO is changed to CO_2, how many electrons must be lost or gained by C? (a) 1 lost (b) 1 gained (c) 2 lost (d) 2 gained ______ 11

DIRECTIONS: Complete the following statements, forming accurate sentences.

12. Any species whose oxidation number becomes more positive is said to be ____________________. 12
13. Any species whose oxidation number becomes more negative is said to be ____________________. 13
14. The oxidation number of a free element is equal to ____________________. 14
15. The oxidation number of any monatomic ion is equal to its ____________________. 15

Name ______________________ Class ____________ Date ____________

Balancing Redox Equations

Section Review 20.2

DIRECTIONS: Write on the line at the right of each statement the letter preceding the word or expression that best completes the statement.

1. In a balanced redox equation, how does the total number of reactant molecules compare to the total number of product molecules? (a) The two numbers are always equal. (b) Reactant molecules are always more numerous. (c) Product molecules are always more numerous. (d) No such relationship always exists between the two numbers. ______ 1
2. In a balanced redox equation, how does the total number of reactant atoms compare to the total number of product atoms? (a) The two numbers are always equal. (b) Reactant atoms are always more numerous. (c) Product atoms are always more numerous. (d) No such relationship always exists between the two numbers. ______ 2
3. The redox equation $Cu^{2+} + 2Fe \rightarrow Cu + 2Fe^{2+}$ is (a) correctly balanced; (b) correctly balanced for numbers of atoms, but not for charge; (c) correctly balanced for charge, but not for numbers of atoms; (d) not balanced for numbers of atoms or charge. ______ 3
4. The redox equation $H_2 + O_2 \rightarrow 2H^+ + O^{2-}$ is (a) correctly balanced; (b) correctly balanced for numbers of atoms, but not for charge; (c) correctly balanced for charge, but not for numbers of atoms; (d) not balanced for numbers of atoms or charge. ______ 4
5. In the oxidation-number method of balancing redox equations, the first step is to (a) adjust coefficients in electronic equations; (b) write a skeleton equation with correct formulas; (c) write two separate electronic equations; (d) place coefficients into the skeleton equation. ______ 5
6. Using the oxidation-number method to balance the redox equation $Cu + HNO_3 \rightarrow Cu(NO_3)_2 + NO + H_2O$, the final coefficients, in order from left to right, should be (a) 1, 4, 1, 1, 2; (b) 3, 8, 3, 2, 4; (c) 2, 6, 2, 1, 2; (d) 1, 2, 2, 1, 1. ______ 6
7. Using the oxidation-number method to balance the redox equation $MnO_4^- + H_2S + H^+ \rightarrow S + Mn^{2+} + H_2O$, the final coefficients, in order from left to right, should be (a) 1, 3, 3, 3, 1, 4; (b) 2, 5, 6, 5, 2, 8; (c) 2, 5, 5, 5, 2, 8; (d) 1, 3, 3, 3, 2, 4. ______ 7
8. Using the ion-electron method to balance the redox equation $HBr + NaMnO_4 \rightarrow NaBr + MnBr_2 + Br_2 + H_2O$, the final coefficients, in order from left to right, should be (a) 8, 2, 2, 2, 5, 8; (b) 4, 1, 1, 1, 3, 2; (c) 16, 2, 2, 2, 3, 4; (d) 2, 16, 10, 2, 8, 5. ______ 8

DIRECTIONS: Fill in the spaces below with the correct equations.

9. Use the oxidation-number method to balance the redox equation
$$MnO_2 + KBr + H_2SO_4 \rightarrow Br_2 + MnSO_4 + K_2SO_4 + H_2O.$$ 9

10. Use the oxidation-number method to balance the redox equation in which manganese(IV) dioxide and hydrochloric acid react to produce manganese(II) chloride, chlorine, and water. 10

11. Use the oxidation-number method to balance the redox equation
$$HSO_3^- + IO_3^- \rightarrow I_2 + SO_4^{2-} + H^+ + H_2O.$$ 11

12. Use the ion-electron method to balance the redox equation
$$Cr_2O_7^{2-} + H^+ + NO_2^- \rightarrow Cr^{3+} + H_2O + NO_3^-.$$ 12

13. Use the ion-electron method to balance the redox equation
$$Br^- + OH^- + N_2O_4 \rightarrow BrO_3^- + H_2O + NO_2^-.$$ 13

Name ______________________ Class ____________ Date ____________

Oxidizing and Reducing Agents

Section Review 20.3

DIRECTIONS: Write on the line at the right of each statement the letter preceding the word or expression that best completes the statement.

1. The most active reducing agent among the elements is (a) cesium; (b) iodine; (c) fluorine; (d) lithium. ______ 1
2. The most active oxidizing agent among the elements is (a) cesium; (b) iodine; (c) fluorine; (d) lithium. ______ 2
3. A process in which a substance acts as both an oxidizing and a reducing agent and oxidizes itself is called (a) electrolysis; (b) auto-oxidation; (c) auto-reduction; (d) double replacement. ______ 3
4. Peroxide ions have the formula (a) O^{2-}; (b) O^{-}; (c) O_2^{-}; (d) O_2^{2-}. ______ 4
5. The bond within a peroxide ion is (a) a double bond; (b) highly stable; (c) somewhat unstable; (d) a triple bond. ______ 5
6. When hydrogen peroxide decomposes, its oxygen is (a) reduced only; (b) oxidized only; (c) both oxidized and reduced; (d) electrolyzed. ______ 6

DIRECTIONS: Write the answer to questions 7–9 on the line to the right, and show your work in the space provided.

7. What is the mass of one equivalent of Fe^{3+} in the reaction $Fe^{3+} + e^{-} \rightarrow Fe^{2+}$? ______ 7

8. What is the mass of one equivalent of Zn in the reaction $Zn + Cu^{2+} \rightarrow Zn^{2+} + Cu$? ______ 8

9. What is the mass of one equivalent of potassium metal in the redox equation in which lead(II) nitrate reacts with potassium metal to produce potassium nitrate and lead metal? ______ 9

DIRECTIONS: In the parentheses at the right of each oxygen-containing substance in the first column, write the letter of the oxidation number for oxygen in the second column. Then indicate, by circling its formula, the substance that is most likely to undergo auto-oxidation.

10. Na_2O	( )	a. +2
11. Na_2O_2	( )	b. +1
12. O_3	( )	c. 0
13. OF_2	( )	d. −1
		e. −2

Name ______________________ Class __________ Date __________

Electrochemistry

Section Review 20.4

DIRECTIONS: Write on the line at the right of each statement the letter preceding the word or expression that best completes the statement.

1. If reactants in a spontaneous energy-releasing redox reaction are in direct contact, energy is released in the form of (a) light; (b) electrical energy; (c) heat; (d) mechanical energy. ______ 1
2. In a dry cell, oxidation of (a) zinc occurs at the negative electrode; (b) manganese occurs at the negative electrode; (c) zinc occurs at the positive electrode; (d) manganese occurs at the positive electrode. ______ 2
3. The transfer of electric charge in the electrolyte between electrodes occurs by means of (a) electron migration; (b) ionization; (c) ion migration; (d) proton migration. ______ 3
4. When one mole of water is decomposed, (a) no energy is absorbed or released; (b) 30.0 kJ of energy is released; (c) 285.9 kJ of energy is released; (d) 285.9 kJ of energy is absorbed. ______ 4
5. Electrolysis solutions containing which of the following produces O_2 at the anode? (a) Br^- (b) I^- (c) Cl^- (d) NO_3^- ______ 5
6. In an electroplating cell, the object to be plated serves as the (a) external circuit; (b) electrolyte; (c) anode; (d) cathode. ______ 6
7. In an electroplating cell, a solution of the salt of the plating metal serves as the (a) external circuit; (b) electrolyte; (c) anode; (d) cathode. ______ 7
8. What is the voltage of the standard automobile battery? (a) 1.5 volts (b) 6 volts (c) 12 volts (d) 50 volts ______ 8
9. When an automobile battery is being charged, (a) heat energy is converted to energy of motion; (b) energy of motion is converted to heat energy; (c) chemical energy is converted to electrical energy; (d) electrical energy is converted to chemical energy. ______ 9
10. The standard electrode that is assigned a potential of 0 volts is a (a) zinc electrode; (b) hydrogen electrode; (c) lead electrode; (d) copper electrode. ______ 10

DIRECTIONS: Write the word "cathode" or "anode" on each line.

11. The electrode at which oxidation occurs. ______________________ 11
12. The electrode at which reduction occurs. ______________________ 12
13. The negative electrode in an electrochemical cell. ______________________ 13
14. The negative electrode in an electrolytic cell. ______________________ 14

15. Describe the rate and direction at which a reaction takes place when E° is positive. ______________________

____

____ 15

Name ______________________ Class ____________ Date ____________

Abundance and Importance of Carbon

Section Review 21.1

DIRECTIONS: Write on the line at the right of each statement the letter preceding the word or expression that best completes the statement.

1. Large deposits of carbon are found in the earth's crust as (a) carbides; (b) dissolved carbon monoxide; (c) carbohydrates; (d) carbonates. ______ 1
2. In terms of abundance of elements in the earth's crust, carbon ranks (a) first; (b) third; (c) sixth; (d) seventeenth. ______ 2
3. Roughly how many carbon compounds have been synthesized in the laboratory? (a) dozens (b) hundreds (c) thousands (d) hundreds of thousands ______ 3
4. The electronic configuration of carbon in the ground state is (a) $1s^22s^22p^2$; (b) $1s^22s^12p^3$; (c) $1s^22s^22p^3$; (d) $1s^22s^22p^6$. ______ 4
5. How many covalent bonds can a carbon atom ordinarily form? (a) 2 (b) 3 (c) 4 (d) 5 ______ 5
6. Carbon atoms join readily with atoms of (a) other elements only; (b) carbon only; (c) both other elements and carbon; (d) only neutral elements. ______ 6
7. Which of the following countries contains the largest diamond mines? (a) South Africa (b) Egypt (c) Brazil (d) the United States ______ 7
8. The actual bonding between atoms in a layer of graphite consists of (a) single bonds only; (b) double bonds only; (c) alternating single and double bonds; (d) bonds that are intermediate in character between single and double. ______ 8
9. Which of the following is NOT an example of amorphous carbon? (a) coke (b) graphite (c) carbon black (d) activated carbon ______ 9
10. The arrangement of atoms in amorphous carbon actually is (a) completely orderly; (b) completely disorderly; (c) partially orderly and partially disorderly, by region; (d) colloidal. ______ 10
11. Most anthracite is found in (a) Pennsylvania; (b) the Rocky Mountain states; (c) the Midwest; (d) California. ______ 11

DIRECTIONS: In the parentheses at the right of each expression in the first column, write the letter of the expression in the second column that is most closely related.

12. partially carbonized plant material ( )
13. absorbs liquids and gases. ( )
14. used to make drugs, dyes, and explosives ( )
15. used as a fuel and as a reducing agent for metal ores ( )

a. activated carbon
b. coal tar
c. peat
d. anthracite
e. coke

Name ______________________ Class __________ Date __________

Organic Compounds

Section Review **21.2**

DIRECTIONS: Write on the line at the right of each statement the letter preceding the word or expression that best completes the statement.

1. Roughly how many organic compounds are known? (a) 400 (b) 4000 (c) 4 million (d) 4 billion _______ 1
2. As the number of carbon atoms in a molecular formula increases, the number of possible isomers (a) increases rapidly; (b) increases slowly; (c) remains the same; (d) decreases. _______ 2
3. Roughly how many new organic compounds are isolated each year? (a) 1 thousand (b) 10 thousand (c) 100 thousand (d) 1 million _______ 3
4. Which of the following is one of the two major divisions of organic compounds? (a) carbonates (b) alkanes (c) hydrocarbons (d) polymers _______ 4
5. Organic compounds are generally classified according to their (a) length; (b) functional groups; (c) mass; (d) reactivity. _______ 5
6. The prefix n in the name n-pentane stands for (a) nitrogen; (b) neo-; (c) number of moles; (d) normal. _______ 6
7. How many structural isomers of CH_4 exist? (a) 1 (b) 2 (c) 3 (d) 5 _______ 7
8. How many structural isomers of C_6H_{14} exist? (a) 1 (b) 2 (c) 3 (d) 5 _______ 8
9. The majority of which of the following tend to be involved in reactions the identity of whose products is not altered by changed reaction conditions? (a) organic compounds only (b) inorganic compounds only (c) both organic compounds and inorganic compounds (d) neither organic compounds nor inorganic compounds _______ 9
10. The laws of chemistry apply to the majority of (a) organic compounds only; (b) inorganic compounds only; (c) both organic compounds and inorganic compounds; (d) undiscovered elements. _______ 10
11. The majority of which of the following exist in molecular form? (a) organic compounds (b) inorganic compounds (c) amphoteric compounds (d) noble gases _______ 11

DIRECTIONS: Write on the line at the right of each statement the word or expression that best completes the meaning when substituted for the corresponding number.

12. The element __(12)__ is contained in all organic compounds. __________ 12
13. Compounds called __(13)__ are compounds that have the same molecular formula but different structures. __________ 13
14. A formula that shows the numbers and types of atoms present in a molecule and that shows how they are bonded to each other is called a(n) __(14)__ formula. __________ 14
15. Most organic compounds are of __(15)__ solubility in water and tend to be involved in reactions that proceed at low rates. __________ 15

Name ______________________ Class ____________ Date ____________

Hydrocarbons

Section Review 21.3

DIRECTIONS: Write on the line at the right of each statement the letter preceding the word or expression that best completes the statement.

1. Hydrocarbons are grouped into several different series mainly on the basis of the (a) number of carbon atoms; (b) isotope of carbon involved; (c) type of bonding between carbon atoms; (d) mass of the compounds. ______ 1
2. Which of the following is an alkane? (a) propyne (b) propane (c) propene (d) propyl bromide ______ 2
3. The structural formula for n-butane is ______ 3

```
    H  H  H             H  H  H               H  H  H  H             H  H  H  H
    |  |  |             |  |  |               |  |  |  |             |  |  |  |
(a) H-C--C--C-H     (b) H-C--C--C-H       (c) H-C--C--C--C-H     (d) H-C--C--C--C-H
    |  |  |             |  |  |               |  |  |  |             |  |  |  |
    H  H  H             H  |  H               H  H  H  H             H  |  H  H
                           |                                            |
                         H-C-H                                        H-C-H
                           |                                            |
                           H                                            H
```

4. The structural formula for cyclohexane is ______ 4

(a) 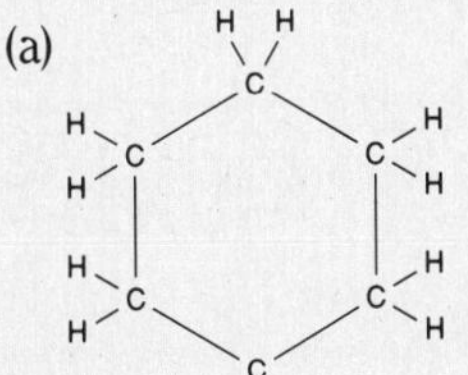

(b) 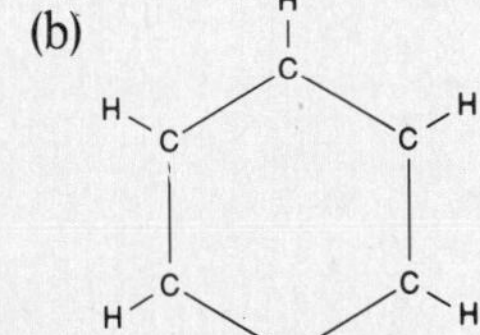

(c) 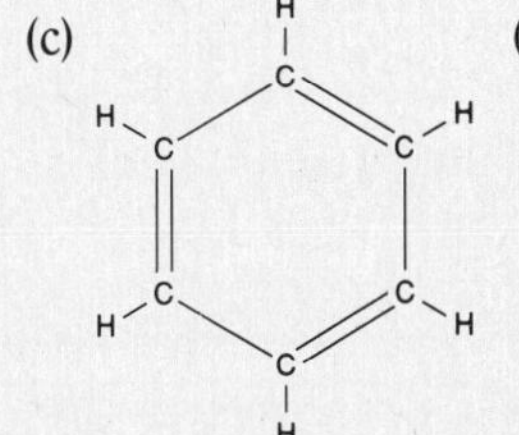

(d) 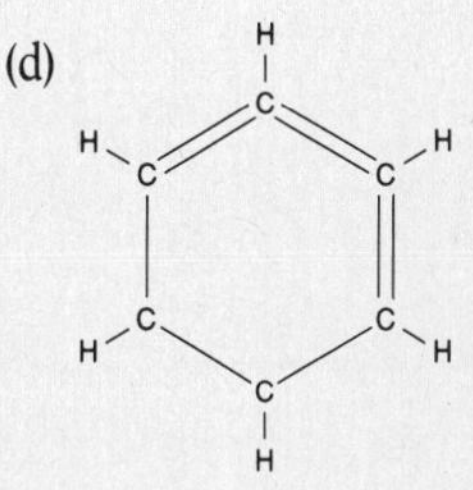

5. What is the name of the compound whose structural formula is

```
H-C=C-H
  |  |
  H  H
```

? (a) ethane (b) ethene (c) ethyne (d) ethadiene ______ 5

6. What is the name of the compound whose structural formula is

```
  H  H  H  H  H  H
  |  |  |  |  |  |
H-C--C--C--C--C--C-H
  |  |  |  |  |  |
  H  |  H  |  H  H
     |     |
   H-C-HH-C-H
     |     |
     H     H
```

? (a) 2,4-dimethylheptane (b) 2,4-diethylhexane (c) 1,4-dimethylhexane (d) 2,4-dimethylhexane ______ 6

7. The hydrocarbon commercially produced in greatest quantity in the United States is (a) ethane; (b) methane; (c) ethyne; (d) ethene. ______ 7
8. The commercial method of producing alkenes involves doing which of the following to petroleum? (a) distilling (b) cracking (c) filtering (d) burning ______ 8
9. A reaction in which gaseous alkanes and alkenes are combined is called (a) substitution; (b) addition; (c) combustion; (d) alkylation. ______ 9
10. The Friedel-Crafts reaction results in (a) hydrogenation of alkenes; (b) halogenation of alkanes; (c) alkylation of benzene; (d) halogenation of benzene. ______ 10
11. Substitution reactions, rather than addition reactions, are most characteristic of (a) alkanes; (b) alkynes; (c) alkadienes; (d) alkenes. ______ 11

DIRECTIONS: Write the names of the following substances on the lines provided.

12. (benzene ring) ______________________ 12

13. (benzene ring with CH_3) ______________________ 13

Name ______________________ Class __________ Date __________

Hydrocarbons and Polymers

Section Review 21.4

DIRECTIONS: Write on the line at the right of each statement the letter preceding the word or expression that best completes the statement.

1. Petroleum is generally a (a) pure substance; (b) mixture of several hydrocarbons; (c) complex mixture of many hydrocarbons; (d) mixture of inorganic substances. ______ 1
2. Crude oil is a (a) simple mixture whose composition is the same in all locations; (b) simple mixture whose composition varies greatly from place to place; (c) complex mixture whose composition is the same in all locations; (d) complex mixture whose composition varies greatly from place to place. ______ 2
3. Natural gas and petroleum were formed (a) hundreds of years ago; (b) several thousand years ago; (c) tens of thousands of years ago; (d) millions of years ago. ______ 3
4. The demand for petroleum will probably exceed available supplies in roughly (a) 20 years; (b) 50 years; (c) 100 years; (d) 500 years. ______ 4
5. Which of the following is obtained as a residue during the refining of petroleum? (a) petroleum ether (b) petroleum coke (c) fuel oil (d) kerosene ______ 5
6. Which of the following is generally used as a solvent or dry-cleaning material rather than as a fuel? (a) petroleum ether (b) kerosene (c) diesel oil (d) gasoline ______ 6
7. Roughly what percentage of the energy used in the United States comes from natural gas and petroleum? (a) 15% (b) 23% (c) 46% (d) 73% ______ 7
8. Which of the following processes is economical in the United States? (a) Fisher-Tropsch only (b) Bergius only (c) both Fischer-Tropsch and Bergius (d) neither Fischer-Tropsch nor Bergius ______ 8
9. A monomer of rubber has the formula (a) C_3H_6; (b) C_4H_8; (c) C_5H_8; (d) C_5H_{10}. ______ 9
10. Neoprene may be used instead of rubber because neoprene is less affected by (a) oxygen; (b) oils and greases; (c) sulfur; (d) heat. ______ 10
11. SBR is used instead of the natural material for which it is a substitute because SBR (a) burns hotter; (b) resists wear better; (c) is less poisonous; (d) is higher-melting. ______ 11

DIRECTIONS: Complete the following statements, forming accurate sentences.

12. The hydrocarbons in natural gas are separated through the process of ______________________. 12
13. One process used to convert coal to fuel gases and liquids is the process called ______________________ ______________________. 13
14. A heating process by which rubber is made more elastic and less sticky, is called ______________________ ______________________. 14
15. Neoprene and SBR are types of synthetic ______________________. 15

Name ______________________ Class __________ Date __________

Alcohols

Section Review 22.1

DIRECTIONS: Write on the line at the right of each statement the letter preceding the word or expression that best completes the statement.

1. The functional group –OH is called a (a) hydroxyl unit; (b) base unit; (c) carboxyl group; (d) hydroxide ion. _______ 1
2. What type of alcohol is isopropyl alcohol? (a) primary (b) secondary (c) tertiary (d) quaternary _______ 2
3. Water solutions of alcohols are generally (a) acidic; (b) slightly basic; (c) neutral; (d) strongly basic. _______ 3
4. Compared to the boiling points of their corresponding hydrocarbons, the boiling points of alcohols are generally (a) higher; (b) the same; (c) slightly lower; (d) much lower. _______ 4
5. The burning of an alcohol in air yields (a) an aldehyde; (b) an ether; (c) a carboxylic acid; (d) carbon dioxide and water. _______ 5
6. The correct equation for the reaction of propene with water in the presence of the catalyst H_2SO_4 is (a) $C_3H_6 + H_2SO_4 \rightarrow C_3H_6SO_4 + H_2$; (b) $C_3H_6 + H_2O \rightarrow C_3H_7OH$; (c) $C_2H_4 + H_2O \rightarrow C_2H_5OH$; (d) $C_3H_6 + 6H_2O \rightarrow 3CO_2 + 9H_2$. _______ 6
7. The reaction between n-butyl chloride and hydroxide ion produces (a) n-butanol; (b) butanoic acid; (c) n-butanal; (d) acetone. _______ 7
8. Which of the following alkyl halides will react with hydroxide ion to produce methanol? (a) ethyl chloride (b) propyl chloride (c) methyl chloride (d) none of the above _______ 8
9. Which of the following is highly poisonous, and is used as a solvent and in gas tank de-icers? (a) ethanol (b) ethylene glycol (c) glycerol (d) methanol _______ 9
10. Which of the following is produced by catalytic hydrogenation of carbon monoxide under pressure? (a) ethanol (b) ethylene glycol (c) glycerol (d) methanol _______ 10
11. Which of the following is synthesized from propene? (a) ethanol (b) ethylene glycol (c) glycerol (d) methanol _______ 11

DIRECTIONS: Write on the line at the right of each statement the word or expression that best completes the meaning when substituted for the corresponding number.

12. The systematic name __(12)__ stands for the one-carbon alcohol. __________ 12
13. A type of compound called a(n) __(13)__ is produced by the reaction of an alcohol with a concentrated hydrogen halide solution. __________ 13
14. When ethene reacts with water in the presence of the catalyst H_3PO_4, __(14)__ is produced. __________ 14
15. Glycerol contains __(15)__ hydroxyl group(s). __________ 15

Name ______________________ Class __________ Date __________

Halocarbons

Section Review **22.2**

DIRECTIONS: Write on the line at the right of each statement the letter preceding the word or expression that best completes the statement.

1. The structural formula of 1,1,3-triiodobutane is

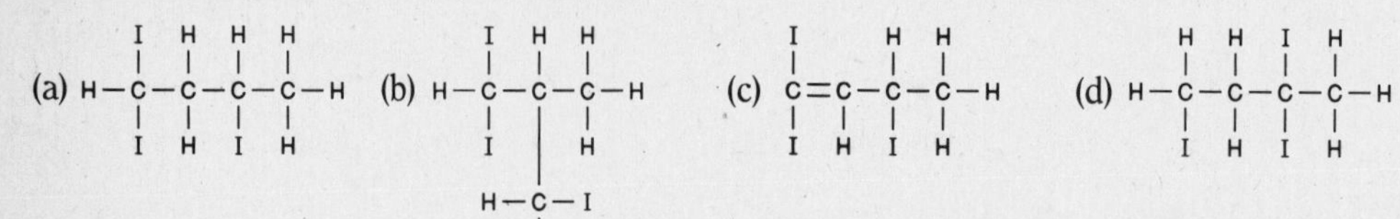

_______ 1

2. What is the name of the compound whose structural formula is

Cl
|
Cl—C—F ?
|
Cl

(a) methyl fluoride (b) 1,1,1-trichloro-2-fluoromethane
(c) trichlorofluoromethane (d) 3-chloro-1-fluoromethane _______ 2

3. Halocarbons can be prepared by direct substitution reactions of halogens with (a) alkanes; (b) alkynes; (c) alkenes; (d) alkalis. _______ 3

4. Halocarbons can be prepared by addition reactions between halogens and (a) alkanes only; (b) alkenes only; (c) hydrocarbons only; (d) both alkanes and alkenes. _______ 4

5. How many hydrogen atoms in CH_4 can be replaced by halogen atoms through direct halogenation? (a) one (b) two (c) three (d) four _______ 5

6. When an alkane and a halogen react, which of the following is one of the products? (a) water (b) carbon (c) a hydrogen halide (d) an alkene _______ 6

7. The reaction of hydroxyl ion with a halocarbon is what type of reaction? (a) addition (b) substitution (c) decomposition (d) single replacement _______ 7

8. What are the products of a reaction between chloroethane and hydroxide ion? (a) ethanol and chloride ion (b) hydrochloroethane (c) chloroethane and water (d) ethanol and chlorine _______ 8

9. What is produced when chloromethane and benzene react in the presence of aluminum chloride? (a) phenyl chloride (b) methyl chloride (c) methanol (d) toluene _______ 9

10. Which of the following is used for dry-cleaning and is poisonous? (a) dichlorodifluoromethane (b) Teflon (c) tetrachloromethane (d) polyvinylchloride _______ 10

11. Which of the following was until recently used as a propellant in aerosol products? (a) dichlorodifluoromethane (b) Teflon (c) tetrachloromethane (d) polyvinylchloride _______ 11

DIRECTIONS: Complete the following statements, forming accurate sentences.

12. Alkanes in which one or more atoms of fluorine, chlorine, bromine, or iodine are substituted for hydrogen atoms are called ______________________________. 12

13. In the Friedel-Crafts reaction, an alkyl halide typically reacts with ______________________________. 13

14. A halocarbon used in plastic pipe is ______________________________. 14

15. Dichlorodifluoromethane is an example of a set of compounds used in refrigeration and called ______________________________. 15

Name ______________________ Class __________ Date __________

Ethers

Section Review 22.3

DIRECTIONS: Write on the line at the right of each statement the letter preceding the word or expression that best completes the statement.

1. The general formula for ethers is (a) ROR′; (b) RCOOH′; (c) RCOOR′; (d) RCHO. ______ 1
2. What is the structural formula of dimethyl ether?

```
     H  H      H  H  H           H                     H     H               H     H
     |  |      |  |  |            \                    |     |               |     |
(a) H-C--C--O--C--C--C-H   (b) H-C--O   H    (c) H-C--O--C-H    (d) H-C--C--C-H
     |  |      |  |  |            |    \ /             |     |               |  ||  |
     H  H      H  H  H            O----C               H     H               H  O   H
                                       |
                                       H
```

______ 2

3. What is the structural formula of propyl ethyl ether?

```
     H  H  H     H  H                  H  H  H     H  H
     |  |  |     |  |                  |  |  |     |  |
(a) H-C--C--C--O--C--C-H         (b) H-C--C--C--C--C--C-H
     |  |  |     |  |                  |  |  |  ||  |  |
     H  H  H     H  H                  H  H  H  O   H  H

     H  H   O                          H  H  H  H  H  H
     |  |  //                          |  |  |  |  |  |
(c) H-C--C--C      H  H          (d) H-C--C--C--C--C--C-H
     |  |   \      |  |                |  |  |  |  |  |
     H  H    O--C--C-H                 H  H  H  O  H  H
                |  |                             \
                H  H                              H
```

______ 3

4. The dehydration of ethanol with sulfuric acid yields (a) ethane; (b) diethyl ether; (c) dimethyl ether; (d) diethyl ketone. ______ 4
5. An ether whose R and R′ groups are different is called a(n) (a) alcohol; (b) mixed ether; (c) double ether; (d) ester. ______ 5
6. In the Williamson synthesis, the reactants are a (a) halocarbon and the sodium salt of an alcohol; (b) halocarbon and a hydroxide; (c) halogen and the sodium salt of an alcohol; (d) halocarbon and an ether. ______ 6
7. A product of a reaction between sodium ethoxide and bromomethane is (a) bromoethane; (b) dimethyl ether; (c) methyl ethyl ether; (d) methyl ethyl ketone. ______ 7
8. Ethers are (a) highly flammable; (b) moderately flammable; (c) slightly flammable; (d) nonflammable. ______ 8
9. In water, ethers are (a) completely miscible; (b) very soluble; (c) slightly soluble; (d) insoluble. ______ 9
10. Ethers are mainly used as (a) medicines; (b) fuels; (c) antiseptics; (d) solvents. ______ 10
11. An ether used often in gasoline to increase octane rating is (a) diethyl ether; (b) dimethyl ether; (c) methyl ethyl ether; (d) methyl butyl ether. ______ 11

DIRECTIONS: Complete the following statements, forming accurate sentences.

12. Compounds in which two hydrocarbon groups are bonded to the same oxygen atom are called ______________________________. 12
13. The Williamson synthesis produces ______________________________. 13
14. The ether that was formerly used frequently as an anaesthetic is ______________________________. 14
15. When treated with hot concentrated mineral acids, ethers ______________________________. 15

Name ______________________ Class __________ Date __________

Aldehydes and Ketones

Section Review 22.4

DIRECTIONS: Write on the line at the right of each statement the letter preceding the word or expression that best completes the statement.

1. The functional group of an aldehyde is called the (a) formyl group; (b) alkyl group; (c) carboxyl group; (d) hydroxyl group. ______ 1
2. The systematic names of ketones end in (a) -one; (b) -al; (c) -ol; (d) -ane. ______ 2
3. The common name for dimethyl ketone is (a) acetaldehyde; (b) formaldehyde; (c) acetone; (d) methyl ether. ______ 3
4. Methanol vapor reacts with heated copper to produce (a) ethanol; (b) formaldehyde; (c) ethanal; (d) dimethyl ketone. ______ 4
5. The oxidation of ethanol produces (a) formaldehyde; (b) acetaldehyde; (c) acetone; (d) diethyl ketone. ______ 5
6. A compound that contains a carbonyl group to which two carbon groups are bonded is called a(n) (a) alcohol; (b) aldehyde; (c) ether; (d) ketone. ______ 6
7. A compound that contains a carbonyl group to which a hydrogen atom is bonded is called a(n) (a) alcohol; (b) aldehyde; (c) ether; (d) ketone. ______ 7
8. Compared to aldehydes, ketones are (a) much more readily oxidized; (b) slightly more readily oxidized; (c) oxidized equally readily; (d) less readily oxidized. ______ 8
9. A compound that is used as a solvent and to saturate the asbestos filling in ethyne storage tanks is (a) diethyl ketone; (b) methyl ethyl ketone; (c) acetone; (d) methanal. ______ 9
10. The general methods of preparing aldehydes is (a) reduction of alcohols; (b) oxidation of alcohols; (c) dehydration of alcohols; (d) dehydration of ketones. ______ 10

DIRECTIONS: Complete the following statements, forming accurate sentences.

11. A compound that contains a carbonyl group to which a hydrogen atom is bonded is called a(n) ______________________________. 11
12. The common name for the compound methanal is ______________________________. 12
13. Mild oxidation of aldehydes produces ______________________________. 13
14. Mild oxidation of 2-propanol produces ______________________________. 14

Name ______________________ Class ____________ Date ____________

Carboxylic Acids

Section Review **22.5**

DIRECTIONS: Write on the line at the right of each statement the letter preceding the word or expression that best completes the statement.

1. The structural formula of acetic acid is

```
     H   O               H  H    O                    O               H  H  H    O
     |  //               |  |   //                   //               |  |  |   //
(a) H-C-C          (b) H-C-C-C          (c) H-C           (d) H-C-C-C-C
     |  \                |  |   \                   \                 |  |  |   \
     H   O-H             H  H    O-H                 O-H              H  H  H    O-H
```
_______ 1

2. The acid whose name derives form the fact that it was once obtained from the distillation of ants is (a) formic acid; (b) butyric acid; (c) acetic acid; (d) propionic acid. _______ 2
3. The acid whose name derives from the Latin word for vinegar is (a) formic acid; (b) butyric acid; (c) acetic acid; (d) propionic acid. _______ 3
4. The attractions between carboxylic acid and molecules consist mainly of (a) van der Waals forces; (b) ionic bonds; (c) London forces; (d) hydrogen bonds. _______ 4
5. Compared to the boiling points of alcohols of corresponding molecular mass, the boiling points of carboxylic acids are (a) higher; (b) roughly equal; (c) somewhat lower; (d) much lower. _______ 5
6. Carboxylic acids are commonly prepared by (a) oxidation of secondary alcohols; (b) reduction of secondary alcohols; (c) oxidation of primary alcohols; (d) reduction of primary alcohols. _______ 6
7. Carboxylic acids are commonly prepared by (a) oxidation of aldehydes; (b) reduction of aldehydes; (c) oxidation of ketones; (d) reduction of ketones. _______ 7
8. Catalyzed reaction of carbon monoxide and methanol produces (a) formic aid; (b) acetic acid; (c) propionic acid; (d) acetone. _______ 8
9. The acid used for dyeing, calico printing, removing stains, and bleaching straw and wood is (a) acetic acid; (b) butyric acid; (c) formic acid; (d) oxalic acid. _______ 9
10. It is true that the two carbon-oxygen bonds in formic acid and the two carbon-oxygen bonds in formate ion are (a) equally long; (b) unequal in length; (c) equal in length in formic acid, but unequal in length in formate ion; (d) equal in length in formate ion, but unequal in length in formic acid. _______ 10

DIRECTIONS: Write on the line at the right of each statement the word or expression that best completes the meaning when substituted for the corresponding number.

11. Organic acids contain the __(11)__ group. ____________ 11
12. Carboxylic acids are __(12)__ acids. ____________ 12
13. Reaction of a carboxylic acid and a hydroxide produces __(13)__ and water. ____________ 13
14. Catalytic oxidation of acetaldehyde produces __(14)__ . ____________ 14

Name ______________________ Class ____________ Date ____________

Esters

Section Review 22.6

DIRECTIONS: Write on the line at the right of each statement the letter preceding the word or expression that best completes the statement.

1. The name of the ester that contains an ethyl group combined with an acetate group is (a) acetic ethane; (b) ethyl ethanoate; (c) methyl acetate; (d) acetic ethanoate. ______ 1

2. What is the structural formula of methyl propanoate? ______ 2

(a)

```
  H    O
  |  //H   H   H
H-C-C  |   |   |
  |  \ |   |   |
  H    O-C-C-H
         | | |
         H H H
```

(b)

```
    H H    O
    | |  //H
H-C-C-C    |
    | |  \ |
    H H    C-H
           |
           H
```

(c)

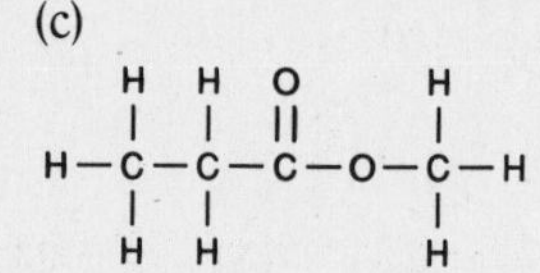

(d)

```
  O     H H H
  ||    | | |
H-C-O-C-C-C-H
        | | |
        H H H
```

3. What is the name of the compound represented by

```
    H H    O
    | |  //   H H H H H
H-C-C-C       | | | | |
    | |  \
    H H    O-C-C-C-C-C-H
             | | | | |
             H H H H H
```

? (a) pentyl propanoate (b) propyl pentanoate (c) pentyl acetate (d) butyl propanoate ______ 3

4. Esters are frequently used (a) as water purifiers; (b) in perfumes and flavorings; (c) as indicators; (d) as electrolytes. ______ 4

5. Fats and oils are (a) alcohols; (b) long-chain organic acids; (c) esters of ethyl alcohol and long-chain organic acids; (d) esters pf glycerol and long-chain organic acids. ______ 5

6. The general method of decomposing an ester into an alcohol and an acid is by (a) esterification; (b) hydrolysis; (C) electrolysis; (d) combustion. ______ 6

7. Reactions between alcohols and esters are (a) reversible and slow to come to equilibrium; (b) reversible and fast to come to equilibrium; (c) irreversible and slow to come to equilibrium; (d) irreversible and fast to come to equilibrium. ______ 7

8. A product of reactions between alcohols and carboxylic acids is (a) hydrogen; (b) oxygen; (c) water; (d) carbon dioxide. ______ 8

9. A catalyst often used in reactions between alcohols and carboxylic acids is (a) aluminum chloride; (b) sulfuric acid; (c) nickel; (d) acetone. ______ 9

10. The hydrolysis of butyl formate in the presence of acids (a) does not occur; (b) produces butyric acid and methyl alcohol; (c) produces butyric acid and formic acid; (d) produces butyl alcohol and formic acid. ______ 10

11. Oils and fats differ in (a) physical phases at room temperature; (b) the number of carbon atoms in the alcohols that produced them; (c) acidity; (d) basic structure. ______ 11

DIRECTIONS: Write on the line at the right of each statement the word or expression that best completes the meaning when substituted for the corresponding number.

12. Compounds called __(12)__ have the general formula RCOOR′. ____________ 12

13. The boiling points of esters are __(13)__ than the boiling points of acids with corresponding molecular mass. ____________ 13

14. The reaction between an alcohol and a carboxylic acid is called a(n) __(14)__ reaction. ____________ 14

15. The reaction called __(15)__ is the reaction in which fats are converted, by hydrolysis, to the salts of fatty acids through the action of a hydroxide. ____________ 15

Name ______________________ Class __________ Date __________

The Chemistry of Life

Section Review 23.1

DIRECTIONS: Write on the line at the right of each statement the letter preceding the word or expression that best completes the statement.

1. Organic compounds contain (a) sulfur; (b) oxygen; (c) carbon; (d) nitrogen. ______ 1
2. The largest molecules all contain (a) carbon; (b) uranium; (c) boron; (d) potassium. ______ 2
3. Which of the following is one of the four most common elements in living things? (a) sulfur (b) oxygen (c) potassium (d) sodium. ______ 3
4. Most of the solutions in living things are basically (a) salt solutions; (b) sugar solutions; (c) alcohol solutions; (d) hydrocarbon solutions. ______ 4
5. Starches and cellulose are (a) carbohydrates; (b) lipids; (c) nucleic acids; (d) proteins. ______ 5
6. Fats and waxes are (a) carbohydrates; (b) lipids; (c) nucleic acids; (d) proteins. ______ 6
7. Macromolecules are built up from (a) polymers; (b) monomers; (c) isomers; (d) allotropes. ______ 7
8. Macromolecules are (a) polymers; (b) monomers; (c) isomers; (d) allotropes. ______ 8
9. Wool, silk, and cotton are (a) polymers; (b) monomers; (c) isomers; (d) allotropes. ______ 9
10. Hydrolysis reactions split polymers by (a) adding water; (b) removing water; (c) adding carbon; (d) removing carbon. ______ 10
11. Which of the following are generally reversible? (a) hydrolysis reactions only (b) condensation reactions only (c) both hydrolysis and condensation reactions (d) double-replacement reactions only ______ 11

DIRECTIONS: Complete the following statements, forming accurate sentences.

12. The study of the reactions that occur in living things is called ______________________ . 12
13. Water's main role in living things is to act as a(n) ______________________ . 13
14. Macromolecules are molecules that are very ______________________ . 14
15. A reaction in which monomers are joined is called a(n) ______________________ . 15

Proteins

Section Review 23.2

DIRECTIONS: Write on the line at the right of each statement the letter preceding the word or expression that best completes the statement.

1. Proteins usually, but not always, contain (a) sulfur; (b) sodium; (c) oxygen; (d) fluorine. ______ 1
2. In what type of reaction are proteins formed? (a) hydrolysis (b) decomposition (c) condensation (d) alkylation ______ 2
3. Roughly how many common types of amino acids exist in organisms? (a) 4 (b) 20 (c) 100 (d) 1000 ______ 3
4. The most common protein chain forms which of the following structures? (a) ring (b) lattice (c) square (d) helix ______ 4
5. Which of the following is a digestive enzyme? (a) trypsin (b) hemoglobin (c) glucose (d) urease ______ 5
6. Which of the following is a protease? (a) trypsin (b) urease (c) pepsin (d) cellulose ______ 6
7. Enzymes bind to their substrates with the help of which of the following? (a) minerals only (b) coenzymes only (c) both minerals and coenzymes (d) ions only ______ 7
8. Which of the following are organic molecules? (a) vitamins only (b) minerals only (c) ions only (d) both vitamins and minerals ______ 8
9. Which of the following is a vitamin? (a) magnesium (b) thiamine (c) cytosine (d) guanine ______ 9
10. Pepsin works best at a pH of approximately (a) 1.5; (b) 3.5; (c) 7.5; (d) 10.5. ______ 10
11. Digestive enzymes that work in the intestine work best at a pH of approximately (a) 1.5; (b) 4; (c) 7; (d) 8. ______ 11

DIRECTIONS: Complete the following statements, forming accurate sentences.

12. Protein monomers are called ______________________. 12
13. Protein molecules that act as catalysts are called ______________________. 13
14. Most enzymes work best at a pH of approximately ______________________. 14
15. As temperature gradually increases, enzyme reaction rates tend to ______________________. 15

Name ______________________ Class __________ Date __________

Carbohydrates

Section Review **23.3**

DIRECTIONS: Write on the line at the right of each statement the letter preceding the word or expression that best completes the statement.

1. Which of the following are carbohydrates? (a) sugars only (b) starches only (c) both sugars and starches (d) proteins only ______ 1
2. The ratio of carbon to hydrogen to oxygen in a monosaccharide is generally (a) 1:2:1; (b) 1:1:2; (c) 2:1:1; (d) 2:2:1. ______ 2
3. Which of the following is a carbohydrate monomer? (a) glucose only (b) sucrose only (c) both glucose and sucrose (d) pepsin only ______ 3
4. Which of the following is a carbohydrate monomer? (a) lactose only (b) cellulose only (c) both lactose and cellulose (d) neither lactose nor cellulose ______ 4
5. A dissolved monosaccharide molecule with five or more carbons takes on which of the following shapes? (a) zigzag line (b) tetrahedron (c) ring (d) helix ______ 5
6. A monosaccharide (a) provides quick energy; (b) stores energy in plants; (c) provides structure in plants; (d) stores energy in animals. ______ 6
7. Starch (a) provides quick energy; (b) stores energy in plants; (c) provides structure in plants; (d) stores energy in animals. ______ 7
8. Which of the following is the main form in which carbohydrates are transported in blood? (a) sucrose (b) glucose (c) glycogen (d) lactose ______ 8
9. Which of the following can be hydrolyzed into glucose? (a) starch only (b) glycogen only (c) both starch and glycogen (d) cellulose only ______ 9
10. Monosaccharides are produced from disaccharides by (a) reduction; (b) hydrolysis; (c) condensation; (d) hydration. ______ 10
11. Polysaccharides are produced from monosaccharides by (a) sucrose; (b) fructose; (c) glucose; (d) glycogen. ______ 11
12. Cotton fibers consist almost entirely of (a) cellulose; (b) glucose; (c) chitin; (d) sucrose. ______ 12

DIRECTIONS: In the parentheses at the right of each word or expression in the first column, write the letter of the expression in the second column that is most closely related.

13. monosaccharide ( )
14. disaccharide ( )
15. polysaccharide ( )

a. sucrose
b. glucose
c. protease
d. glycogen

Name ______________________ Class __________ Date __________

Lipids

Section Review 23.4

DIRECTIONS: Write on the line at the right of each statement the letter preceding the word or expression that best completes the statement.

1. Lipids are (a) organic compounds of high solubility in water; (b) organic compounds of low solubility in water; (c) inorganic compounds of high solubility in water; (d) inorganic compounds of low solubility in water. ______ 1
2. Lipids are highly soluble in (a) ether only; (b) water only; (c) both ether and water; (d) salt water only. ______ 2
3. Lipids are (a) macromolecules composed of monomers; (b) macromolecules not composed of monomers; (c) relatively small molecules composed of monomers; (d) relatively small molecules not composed of monomers. ______ 3
4. Which of the following form a waterproof coating on skin, hair, feathers, and leaves? (a) phospholipids (b) waxes (c) steroids (d) nucleic acids ______ 4
5. Which of the following are the chief lipids in membranes? (a) phospholipids (b) waxes (c) steroids (d) nucleic acids ______ 5
6. Which of the following is a sex hormone? (a) cortisone (b) estrogen (c) epinephrine (d) thyroxin ______ 6
7. Which of the following is a local chemical messenger? (a) thyroxin (b) testosterone (c) histamine (d) epinephrine ______ 7
8. Which of the following is an element in all lipids? (a) nitrogen (b) sulfur (c) potassium (d) oxygen ______ 8
9. Which of the following is an element in all lipids? (a) hydrogen (b) fluorine (c) magnesium (d) iodine ______ 9
10. Which type of bonding produces unsaturated fatty acids? (a) C–C (b) C=C (c) C–O (d) O–H ______ 10

DIRECTIONS: In the parentheses at the right of each expression in the first column, write the letter of the expression in the second column that is most closely related.

11. consist of a hydrocarbon chain with a carboxyl group at one end ( )
12. contain a very long hydrocarbon chain with an alcohol group on it ( )
13. contain four carbon rings ( )
14. the most important structural lipids ( )
15. include fats and oils ( )

a. waxes
b. phospholipids
c. triglycerides
d. fatty acids
e. steroids
f. carbohydrates

Name ______________________ Class ____________ Date ____________

Nucleic Acids

Section Review 23.5

DIRECTIONS: Write on the line at the right of each statement the letter preceding the word or expression that best completes the statement.

1. Which of the following are NOT parts of a nucleotide? (a) phosphate groups (b) amino acids (c) nitrogen bases (d) sugars ______ 1
2. How many carbon atoms are in each sugar in nucleic acids? (a) five (b) six (c) twelve (d) sixteen ______ 2
3. Nucleotides in cells generally each contain how many phosphate groups? (a) 1 to 3 (b) 3 to 5 (c) 6 to 9 (d) 10 to 12 ______ 3
4. DNA strands occur in (a) pairs; (b) threes; (c) fours; (d) fives. ______ 4
5. In DNA, phosphates are linked to (a) A bases; (b) T bases; (c) other phosphates; (d) sugars. ______ 5
6. What holds DNA strands together? (a) base-pairing (b) phosphate-phosphate pairing (c) phosphate-sugar pairing (d) sugar-sugar pairing ______ 6
7. Genetic coding exists basically in sequencing of (a) bases; (b) phosphates; (c) ribose molecules; (d) carbon and oxygen atoms. ______ 7
8. DNA codes for (a) lipid structure; (b) protein structure; (c) carbohydrate structure; (d) steroid structure. ______ 8
9. Different kinds of cells differ in an individual because they contain different (a) DNA; (b) enzymes; (c) sugars; (d) amino acids. ______ 9
10. A string of how many nucleotides codes for an amino acid? (a) 3 (b) 10 (c) 20 (d) 50 ______ 10
11. Which of the following has not yet been achieved? (a) isolation of particular genes (b) identification of amino acid sequences in proteins (c) gene transfers (d) mapping of the entire human genome ______ 11

DIRECTIONS: Complete the following statements, forming accurate sentences.

12. The sugar found in DNA is called ______________________. 12
13. The four bases in DNA are indicated by the letters ______________________. 13
14. A segment of DNA that codes for some particular thing is called a(n) ______________________. 14
15. The deliberate scientific manipulation of DNA is called ______________________. 15

Name ______________________ Class __________ Date __________

Biochemical Pathways

Section Review 23.6

DIRECTIONS: Write on the line at the right of each statement the letter preceding the word or expression that best completes the statement.

1. Energy for biological processes comes from (a) the sun directly; (b) food; (b) the atmosphere; (d) water. ______ 1
2. Metabolism involves (a) only the synthesis of molecules; (b) only the breakdown of molecules; (c) only the extraction and use of energy; (d) the synthesis and breakdown of molecules and the extraction and use of energy. ______ 2
3. A metabolic pathway generally involves (a) energy changes only; (b) a single reaction and energy changes; (c) a series of reactions but no energy changes; (d) a series of reactions and energy changes. ______ 3
4. Glycolysis is an example of a(n) (a) genetic change; (b) metabolic pathway; (c) physical change; (d) inorganic reaction. ______ 4
5. The citric acid cycle converts (a) fat to carbohydrate; (b) carbohydrate to protein; (c) protein to fat; (d) fat to protein. ______ 5
6. ATP is essentially a (a) nitrogen base; (b) sugar; (c) nucleotide; (d) lipid. ______ 6
7. Glycolysis is production of (a) glucose from glycogen; (b) glucose from ATP; (c) ATP from glucose; (d) glycogen from ATP. ______ 7
8. What enzyme catalyzes the first reaction in glycolysis? (a) trypsin (b) hexokinase (c) pyruvate (d) protease ______ 8
9. Most energy in organisms is supplied by hydrolysis of (a) sugars; (b) proteins; (c) hexokinase; (d) ATP. ______ 9

DIRECTIONS: Complete the following statements, forming accurate sentences.

10. All of an organism's biochemical reactions are collectively called ______________________. 10
11. The substrate for glycolysis is ______________________. 11
12. ATP is an abbreviation for ______________________. 12
13. The hydrolysis of a phosphate in ATP causes energy to be ______________________. 13

DIRECTIONS: Write the answers to the following on the lines provided.

14. Describe the general process of metabolism. ______________________

____

____

____

______________________ 14

15. What happens during the process of glycolysis? ______________________

____

____

____

______________________ 15

Name ______________________ Class ____________ Date ____________

The Alkali Metals

Section Review **24.1**

DIRECTIONS: Write on the line at the right of each statement the letter preceding the word or expression that best completes the statement.

1. Which of the following is the chemical symbol for an alkali metal? (a) H (b) Co (c) S (d) Cs ______ 1
2. The alkali metals are (a) hard; (b) nonmalleable; (c) poor conductors of heat; (d) ductile. ______ 2
3. The oxides of alkali elements are (a) acid anhydrides; (b) basic anhydrides; (c) neutral; (d) completely unreactive. ______ 3
4. The unit cell of alkali elements is (a) face-centered; (b) body-centered; (c) amorphous; (d) square. ______ 4
5. How many nearest neighbors does an ion in the unit cell of an alkali element have? (a) one (b) four (c) six (d) eight ______ 5
6. The malleability of alkali elements is explained mainly on the basis of (a) binding force; (b) electron mobility; (c) unit-cell shape; (d) body-centeredness. ______ 6
7. How many electrons do the atoms of alkali elements have in their highest energy level? (a) one (b) two (c) seven (d) eight ______ 7
8. How many electrons do the ions formed most readily by alkali metals have in their outermost energy level? (a) one (b) two (c) seven (d) eight ______ 8
9. The flame color of sodium is (a) red; (b) yellow; (c) magenta; (d) violet. ______ 9
10. The alkali metal that forms a true oxide by direct reaction is (a) cesium; (b) rubidium; (c) lithium; (d) sodium. ______ 10
11. The alkali metal that forms a peroxide by direct reaction is (a) cesium; (b) rubidium; (c) lithium; (d) sodium. ______ 11

DIRECTIONS: In the parentheses at the right of each characteristic in the first column, write the letter of the alkali metal in the second column that is most closely described.

12. does not exist as a stable element	( )	a. sodium
13. most plentiful in nature	( )	b. cesium
14. chemical symbol is K	( )	c. lithium
15. lowest mass of all alkali metals	( )	d. francium
		e. rubidium
		f. potassium

Name ____________________ Class __________ Date __________

The Alkaline–Earth Metals

Section Review 24.2

DIRECTIONS: Write on the line at the right of each statement the word or expression that best completes the meaning when substituted for the corresponding number.

1. Alkaline-earth elements are members of the (a) argon family; (b) calcium family; (c) zinc family; (d) lithium family. ______ 1
2. Alkaline-earth compounds are found mostly in (a) the earth's crust; (b) the atmosphere; (c) the oceans; (d) lakes and rivers. ______ 2
3. Which of the following is NOT a class of alkaline-earth compounds that makes up important natural deposits? (a) sulfides (b) phosphates (c) carbonates (d) silicates ______ 3
4. Which of the following alkaline-earth metals is always radioactive? (a) strontium (b) calcium (c) barium (d) radium ______ 4
5. Compared to the corresponding alkali metals, alkaline-earth metals are (a) denser and softer; (b) denser and harder; (c) less dense and softer; (d) less dense and harder. ______ 5
6. Compared to the corresponding alkali metals, alkaline-earth metals have (a) higher melting and boiling points; (b) lower melting and boiling points; (c) higher melting points and lower boiling points; (d) lower melting points and higher boiling points. ______ 6
7. Compared to the ions of the corresponding alkali metals, the ions of alkaline-earth metals are (a) smaller; (b) equal in size; (c) slightly larger; (d) much larger. ______ 7
8. Covalent bonding exists in (a) CaH_2; (b) SrH_2; (c) BeH_2; (d) BaH_2. ______ 8
9. Beryllium oxide is (a) basic; (b) neutral; (c) acidic; (d) amphoteric. ______ 9
10. Which alkaline-earth metal tends to form peroxides rather than oxides? (a) beryllium (b) magnesium (c) strontium (d) calcium ______ 10
11. Solubility of alkaline-earth hydroxides (a) increases with the metallic ion's size; (b) decreases with the metallic ion's size; (c) remains the same, regardless of the metallic ion's size; (d) does not depend on the metallic ion's size. ______ 11

DIRECTIONS: Complete the following statements, forming accurate sentences.

12. The alkaline-earth element of lowest mass is ____________________. 12
13. The alkaline-earth element of highest mass is ____________________. 13
14. The number of valence electrons in alkaline-earth elements is ____________________. 14
15. A heat-resistant material is said to be ____________________. 15

Name ______________________ Class ____________ Date ____________

Sodium

Section Review 24.3

DIRECTIONS: Write on the line at the right of each statement the letter preceding the word or expression that best completes the statement.

1. Which of the following is a natural source that contains elemental metallic sodium? (a) sea water only (b) salt deposits only (c) both sea water and salt deposits (d) neither sea water nor salt water ______ 1
2. Sodium chloride is present in very large quantities in (a) sea water; (b) iron ore; (c) rain water; (d) tap water. ______ 2
3. Important deposits of sodium nitrate are located in (a) Mexico and Guatemala; (b) Chile and Peru; (c) France and Italy; (d) Great Britain and Ireland. ______ 3
4. Carbonates, sulfates, and borates of sodium are found in large quantities in (a) sea water; (b) dry lake beds; (c) rain water; (d) tap water. ______ 4
5. The scientist who prepared metallic sodium in 1807 was (a) Avogadro; (b) Faraday; (c) Davy; (d) Arrhenius. ______ 5
6. The overall cell reaction for the typical preparation of sodium metal is (a) $NaCl \rightarrow Na^+ + Cl^-$; (b) $2NaCl \rightarrow Na(l) + NaCl_2$; (c) $2NaCl \rightarrow 2Na(l) + Cl_2(g)$; (d) $2NaOH \rightarrow 2Na(l) + H_2(g) + O_2(g)$. ______ 6
7. Sodium is used (a) to make tetraethyl lead only; (b) as a heat-transfer agent only; (c) in sodium-vapor lamps only; (d) to make tetraethyl lead, as a heat-transfer agent, and in sodium vapor lamps. ______ 7
8. Which of the following is mostly obtained by mining in Wyoming or by brine evaporation from Searles Lake, California? (a) $NaCl$ (b) $NaOH$ (c) $NaHCO_3$ (d) Na_2CO_3 ______ 8
9. Which of the following is NOT a raw material in the Solvay process? (a) $NaCl$ (b) Na_2CO_3 (c) limestone (d) coal ______ 9
10. The thermal disintegration of limestone yields (a) CaO and CO_2; (b) NaO and CO_2; (c) Ca and Cl_2; (d) $NaHCO_3$ and H_2O. ______ 10
11. Sodium carbonate is also called (a) soda ash; (b) baking soda; (c) Chile saltpeter; (d) lye. ______ 11

DIRECTIONS: Complete the following statements, forming accurate sentences.

12. Sodium metal is generally prepared today by electrolysis of fused ______________________. 12
13. The apparatus generally used today to prepare sodium metal is called (a)n ______________________. 13
14. When sodium metal is placed into water, the gas produced is ______________________. 14
15. Most sodium carbonate is manufactured by a process called the ______________________. 15

Name ______________________ Class __________ Date __________

The Transition Elements

Section Review 25.1

DIRECTIONS: Write on the line at the right of each statement the letter preceding the word or expression that best completes the statement.

1. How many groups of elements are made up of transition elements? (a) 6 (b) 8 (c) 10 (d) 18 ______ 1
2. Which of the following is NOT a transition element? (a) manganese (b) cobalt (c) lead (d) silver ______ 2
3. The name "transition element" comes from the fact that such elements represent a(n) (a) transition from metals to nonmetals; (b) transition from solid to liquid and gaseous elements; (c) interruption in the regular increase of neutrons in the nucleus; (d) interruption in the regular increase of highest-energy-level electrons. ______ 3
4. How many electrons are in the highest energy level of transition elements? (a) one or two (b) two or three (c) three or four (d) four or more ______ 4
5. Which of the following sublevels may become involved in transition-compound bonding? (a) highest *s* sublevel only (b) highest *p* sublevel only (c) sublevels in highest energy level, and next-to-highest *d* sublevel (d) lowest *s* sublevel only ______ 5
6. Groups of elements headed by which of the following show greater horizontal similarity than vertical similarity? (a) scandium, titanium, and vanadium (b) chromium, manganese and iron (c) iron, cobalt, and nickel (d) nickel, copper, and zinc ______ 6
7. Compared to Group 1 and Group 2 metals, transition metals are generally (a) softer and less brittle; (b) softer and more brittle; (c) harder and less brittle; (d) harder and more brittle. ______ 7
8. The degree of hardness of transition metals is due partially to (a) sharing of *p* electrons; (b) sharing of *d* electrons; (c) ionization; (d) ionic bonding. ______ 8
9. The common oxidation state of scandium is (a) +1; (b) +2; (c) +3; (d) +5. ______ 9
10. The common oxidation states of chromium are (a) +2, +3, and +6; (b) +6 and +7; (c) +1 and +2; (d) +4 and +5. ______ 10
11. In which of their oxidation states are most transition compounds paramagnetic? (a) none (b) all (c) all but the lowest (d) all but the highest ______ 11
12. How many unpaired *d* electrons does a manganese atom have? (a) 0 (b) 1 (c) 5 (d) 7 ______ 12

DIRECTIONS: Write on the line at the right of each statement the word or expression that best completes the meaning when substituted for the corresponding number.

13. The transition elements are __(13)__ conductors of heat and electricity. ________ 13
14. The coordination number of Ag is equal to __(14)__ in $Ag(NH_3)_2^+$. ________ 14
15. __(15)__ is the only metal that is a liquid at room temperature. ________ 15

Name ______________________ Class ____________ Date ____________

Iron, Cobalt and Nickel

Section Review 25.2

DIRECTIONS: Write on the line at the right of each statement the letter preceding the word or expression that best completes the statement.

1. The platinum family includes platinum, (a) nickel and palladium; (b) iridium, and ruthenium; (c) gold, and silver; (d) nickel, and silver. ______ 1
2. The members of the platinum family are called (a) alkali metals; (b) rare-earth elements; (c) actinide elements; (d) noble metals. ______ 2
3. When Fe^{+} is formed from Fe, which electrons are lost? (a) two 4*s* and one 4*p* (b) two 4*s* and one 3*d* (c) one 4*s* and two 3*d* (d) two 4*s* and one 3*p* ______ 3
4. Iron is the (a) most abundant metal; (b) second most abundant metal; (c) third most abundant metal; (d) fourth most abundant metal. ______ 4
5. It is likely that the earth's core consists largely of (a) cobalt; (b) magnesium; (c) aluminum; (d) iron. ______ 5
6. The reducing agent that acts upon iron oxide in a blast furnace is (a) O_2; (b) CO_2; (c) $CaCO_3$; (d) CO. ______ 6
7. What effect do phosphorus and sulfur impurities have on iron? (a) They both make iron brittle at low temperatures. (b) They both make iron brittle at high temperatures. (c) Phosphorus makes iron brittle at low temperatures, and sulfur makes iron brittle at high temperatures. (d) Sulfur makes iron brittle at low temperatures, and phosphorus makes iron brittle at high temperatures. ______ 7
8. Conversion of iron to steel essentially involves removal of impurities by (a) oxidation; (b) reduction; (c) boiling; (d) condensation. ______ 8
9. The most useful compound of iron in the +3 oxidation state is a hydrate of (a) $FeCl_2$; (b) $FeCl_3$; (c) $Fe_2(SO_4)_3$; (d) $FeSO_4$. ______ 9
10. In boiling water, Fe^{3+} ion (a) does not hydrolyze; (b) hydrolyzes to produce OH^- and FeH^{4+} ions; (c) hydrolyzes to produce H_3O^+ and $Fe(OH)_3$; (d) changes to Fe^{2+} ion. ______ 10
11. If chlorine is added to a solution formed by addition of excess cyanide to iron(II) solution, which of the following is formed? (a) $Fe(CN_2)$ (b) $Fe(CN)_4^{2-}$ (c) $Fe(CN)_6^{3-}$ (d) $Fe(CN)_6^{4-}$ ______ 11

DIRECTIONS: In the parentheses at the right of each word or expression in the first column, write the letter of the expression in the second column that is most closely related.

12. carbon ( )
13. earthy impurities ( )
14. material added to cause mineral impurities to melt more readily ( )
15. iron alloy ( )

a. flux
b. flue gas
c. slag
d. steel
e. coke

Name ______________________ Class __________ Date __________

Copper, Silver and Gold

Section Review 25.3

DIRECTIONS: Write on the line at the right of each statement the letter preceding the word or expression that best completes the statement.

1. How many electrons do atoms in the copper family have in their outermost *s* sublevel? (a) none (b) one (c) two (d) three _______ 1
2. What are the common oxidation states for silver? (a) +1 only (b) +1 and +3 (c) +1 and +2 (d) +2 and +3 _______ 2
3. What are the common oxidation states for gold? (a) +1 only (b) +1 and +3 (c) +1 and +2 (d) +2 and +3 _______ 3
4. Which of the following is a green copper ore? (a) azurite (b) hematite (c) taconite (d) malachite _______ 4
5. Most copper is produced from (a) carbonate ores; (b) sulfide ores; (c) chloride ores; (d) sulfate ores. _______ 5
6. Carbonate ores of copper are usually made to produce copper metal by (a) heating in a furnace; (b) precipitation; (c) electrolysis; (d) dissolving. _______ 6
7. Sulfide ores of copper are usually made to produce copper metal by (a) heating in a furnace; (b) precipitation; (c) electrolysis; (d) dissolving. _______ 7
8. One of the most important uses of copper is in making (a) jewelry; (b) electrical wire; (c) coins; (d) superconductors. _______ 8
9. Copper is used in making all of the following EXCEPT (a) plumbing pipes; (b) transmission lines; (c) electrical insulators; (d) roofing and gutters. _______ 9
10. Which of the following is an alloy of copper? (a) brass only (b) bronze only (c) silver only (d) both brass and bronze _______ 10
11. Compared to pure copper, common alloys of copper are usually (a) more expensive; (b) less strong; (c) redder in color; (d) easier to work with. _______ 11

DIRECTIONS: Write the answers to the following on the lines provided. Where appropriate, make complete statements.

12. The two members of the copper family, other than copper itself, are ______________________. 12
13. Bronze is an alloy of ______________________. 13
14. Most copper compounds color a Bunsen flame ______________________. 14
15. Copper reacts only with what type of acids? ______________________ 15

Name ______________________ Class ____________ Date ____________

Introduction to Aluminum and Metalloids

Section Review 26.1

DIRECTIONS: Write on the line at the right of each statement the letter preceding the word or expression that best completes the statement.

1. Which of the following is a metalloid? (a) tellurium (b) tungsten (c) technetium (d) thorium ______ 1
2. Which of the following is NOT a metalloid? (a) aluminum (b) arsenic (c) antimony (d) astatine ______ 2
3. Which of the following is NOT a metalloid? (a) silicon (b) polonium (c) germanium (d) gallium ______ 3
4. What pattern do the metalloids form in the periodic table? (a) a horizontal region (b) a vertical region (c) a diagonal region slanting from upper center toward lower right (d) a diagonal region slanting from lower center toward upper right ______ 4
5. Which of the following is true of the oxide and the hydride of aluminum? (a) The oxide is covalent and the hydroxide is ionic. (b) The oxide is ionic and the hydroxide is covalent. (c) Both the oxide and the hydroxide are covalent. (d) Both the oxide and the hydroxide are ionic. ______ 5
6. When aluminum is heated in air, it is (a) resistant to oxidation; (b) easily oxidized; (c) easily ionized; (d) electrolyzed. ______ 6
7. What is produced when aluminum is placed in aqueous sodium hydroxide? (a) There is no reaction. (b) $NaAl(OH)_4$ and H_2 (c) $Al(OH)_3$ and H_2O (d) AlH_3 and O_2 ______ 7
8. How many metalloids do NOT form compounds in which they have a positive oxidation state? (a) 0 (b) 1 (c) 3 (d) 5 ______ 8
9. Which of the following can form compounds in which it has a negative oxidation state? (a) boron (b) aluminum (c) arsenic (d) polonium ______ 9
10. What is the approximate range of melting points for metalloids? (a) $-250°C$ to $-20°C$ (b) $-20°C$ to $20°C$ (c) $20°C$ to $250°C$ (d) $250°C$ to $2300°C$ ______ 10
11. What is the approximate range of atomic radii, in angstroms, for metalloids? (a) 0.08 to 0.1 (b) 0.1 to 1.0 (c) 0.8 to 1.5 (d) 1.5 to 3.5 ______ 11

DIRECTIONS: Complete the following statements, forming accurate sentences.

12. Another name for metalloids is ______________________. 12
13. In the periodic table, metalloids occur in Groups ______________________. 13
14. The formula of the aluminate ion is ______________________. 14
15. The metalloid that has the smallest atomic radius is ______________________. 15

Name ______________________ Class ____________ Date ____________

Aluminum

Section Review 26.2

DIRECTIONS: Write on the line at the right of each statement the letter preceding the word or expression that best completes the statement.

1. Aluminum is a (a) low-density metal; (b) high-density metal; (c) low-density metalloid; (d) high-density metalloid. ________ 1
2. Which of the following is added to aluminum in small amounts to strengthen it and make it more resistant to corrosion? (a) lead (b) sulfur (c) potassium (d) manganese ________ 2
3. During electrolytic extraction of aluminum, which of the following represents what occurs at the negative electrode? (a) $4Al \rightarrow 4Al^{3+} + 12e^-$ (b) $4Al^{3+} + 12e^- \rightarrow Al$ (c) $6O^{2-} \rightarrow 3O_2 + 12e^-$ (d) $12e^- + 3O_2 \rightarrow 6O^{2-}$ ________ 3
4. How does aluminum rank as a conductor of electricity? (a) first over all metals (b) second, after silver (c) third, after silver and copper (d) fourth, after silver, copper, and gold ________ 4
5. The surface of a sample of aluminum is normally made of (a) pure aluminum; (b) aluminum oxide; (c) aluminum nitride; (d) aluminum carbonate. ________ 5
6. Aluminum reacts readily to liberate hydrogen from (a) both nitric acid and hydrochloric acid; (b) neither nitric acid nor hydrochloric acid; (c) nitric acid but not hydrochloric acid; (d) hydrochloric acid but not nitric acid. ________ 6
7. The net chemical equation for the typical thermite reaction is (a) $Al_2O_3 + Fe_2O_3 \rightarrow 2Al + 2Fe + 3O_2$; (b) $Al_2O_3 + Fe \rightarrow 2Al + Fe_2O_3$; (c) $2Al + 2Fe + 3O_2 \rightarrow Al_2O_3 + Fe_2O_3$; (d) $2Al + Fe_2O_3 \rightarrow 2Fe + Al_2O_3$. ________ 7
8. The thermite reaction is used for all of the following EXCEPT to (a) produce small quantities of carbon-free metal; (b) reduce metallic oxides that are not readily reduced by carbon; (c) reduce cheaper metals on a large scale; (d) make repairs by welding. ________ 8
9. In the typical thermite reaction, the reducing agent is (a) Al; (b) Al_2O_3; (c) Fe; (d) Fe_2O_3. ________ 9
10. Corundum and emery are compounds of aluminum and (a) carbon; (b) oxygen; (c) nitrogen; (d) sulfur. ________ 10
11. The color of sapphires results from the presence of trace quantities of (a) titanium; (b) vanadium; (c) manganese; (d) chromium. ________ 11

DIRECTIONS: Write on the line at the right of each statement the word or expression that best completes the meaning when substituted for the corresponding number.

12. The ore __(12)__ is the major source of aluminum. ____________ 12
13. The highly exothermic reaction involving aluminum and compounds of other metals is called the __(13)__ reaction. ____________ 13
14. A(n) __(14)__ is the name for any stiff material that contains embedded crystalline strands. ____________ 14
15. The color of rubies results from the presence of trace quantities of __(15)__ in alumina. ____________ 15

Name ______________________ Class ____________ Date ____________

Representative Metalloids

Section Review 26.3

DIRECTIONS: Write on the line at the right of each statement the letter preceding the word or expression that best completes the statement.

1. Silicon is a (a) good conductor at all temperatures; (b) semiconductor whose conductivity decreases with increasing temperature; (c) semiconductor whose conductivity increases with increasing temperature; (d) nonconductor at all temperatures. ______ 1
2. How many valence electrons does silicon have? (1) one (2) two (3) three (4) four. ______ 2
3. What percentage of the material in the earth's crust is composed of silicon dioxide and silicates? (a) 7% (b) 27% (c) 47% (d) 87% ______ 3
4. Atoms of silicon have fairly high ionization energy and (a) high electronegativity; (b) low electronegativity; (c) electronegativity of 100; (d) zero electronegativity. ______ 4
5. Manufacture of which of the following is NOT a commercial use for silicon? (a) alloys (b) conducting wires (c) solar cells (d) transistors ______ 5
6. The chemical symbol for arsenic is (a) As; (b) Ar; (c) Sb; (d) Sn. ______ 6
7. When heated in moist air, arsenic (a) does not react; (b) forms a hydrate; (c) forms a nitride; (d) forms an oxide. ______ 7
8. The chemical formula for antimony is (a) As; (b) An; (c) Sb; (d) Sn. ______ 8
9. Antimony and stibnite have been used since antiquity for making (a) paper; (b) clothing; (c) building materials; (d) cosmetics. ______ 9
10. An alloy of antimony used to reduce friction between the surfaces of moving parts in machinery also contains (a) lead; (b) arsenic; (c) iron; (d) silicon. ______ 10

DIRECTIONS: In the parentheses at the right of each element name in the first column, write the letter of the formula in the second column that is most closely related.

11. stibnite	( )	a. $FeAsS$
12. arsenopyrite	( )	b. As_2S_3
13. andetite	( )	c. $Ca_2B_6O_{11}\cdot 5H_2O$
14. quartz	( )	d. SiO_2
15. orpiment	( )	e. Sb_2S_3
		f. As_2O_3

Name ______________________ Class __________ Date __________

Elemental Sulfur

Section Review **27.1**

DIRECTIONS: Write on the line at the right of each statement the letter preceding the word or expression that best completes the statement.

1. Sulfur commonly occurs in nature (a) as the free element only; (b) combined only; (c) combined with water only; (d) both as the free element and combined. _______ 1
2. In what phase is elemental sulfur generally removed from deposits? (a) solid; (b) liquid; (c) gas; (d) plasma. _______ 2
3. A large fraction of the sulfur produced in the United States is obtained from which of the following compounds? (a) FeS_2 (b) CuS (c) MgS (d) PbS_2 _______ 3
4. Sulfur combines directly with (a) all metals; (b) nearly all metals; (c) all nonmetals; (d) no metals. _______ 4
5. Sulfur is used in the vulcanization of (a) rubber; (b) plastics; (c) steel; (d) wood. _______ 5
6. The common oxidation numbers of sulfur in its compounds with oxygen are: (a) -6 and -4; (b) -4 and -2; (c) $+2$ and $+4$; (d) $+4$ and $+6$. _______ 6
7. Combined sulfur generally occurs in the form of (a) sulfur dioxide; (b) sulfur trioxide; (c) sulfites and hydrogen sulfites; (d) sulfides and sulfates. _______ 7
8. The making of which of the following is NOT a common use of sulfur? (a) dyes (b) medicines (c) plant foods (d) fungicides _______ 8

DIRECTIONS: In the parentheses at the right of each expression in the first column, write the letter of the expression in the second column that is most closely related.

9. dark red, viscous liquid	( )	a. alpha-sulfur
10. straw-yellow, highly fluid liquid	( )	b. beta-sulfur
		c. lambda-sulfur
11. rhombic crystal	( )	d. mu-sulfur
12. monoclinic crystal	( )	e. amorphous sulfur

DIRECTIONS: Complete the following statements, forming accurate sentences.

13. A United States state in which large deposits of sulfur are found is ______________________. 13
14. A solvent in which common sulfur is very soluble is ______________________. 14
15. Most sulfur is used to make the compound ______________________. 15

Name ______________________ Class __________ Date __________

Important Compounds of Sulfur

Section Review 27.2

DIRECTIONS: Write on the line at the right of each statement the letter preceding the word or expression that best completes the statement.

1. Sulfur dioxide occurs in significant quantities in all of the following EXCEPT (a) volcanic gas; (b) the products of the burning of natural gas; (c) mineral waters; (d) the products of the burning of coal. ______ 1
2. The roasting of sulfide ores directly produces (a) sulfur; (b) hydrogen sulfide; (c) sulfur dioxide; (d) sulfur trioxide. ______ 2
3. Laboratory preparation of sulfur dioxide often involves the reduction of (a) sulfur; (b) sulfur trioxide; (c) sulfurous acid; (d) sulfuric acid. ______ 3
4. Sulfur dioxide, when oxidized and combined with water, forms (a) hydrosulfuric acid; (b) sulfuric acid; (c) sulfurous acid; (d) sulfur hydroxide. ______ 4
5. Which of the following is commonly used to preserve dried fruits? (a) sulfiting agents (b) sulfuric acid (c) sulfur dioxide (d) sulfur trioxide ______ 5
6. The first step in the usual method of producing sulfuric acid is the production of (a) SO_2; (b) H_2S; (c) H_2SO_3; (d) Na_2SO_4. ______ 6
7. The sulfur trioxide produced in the manufacture of H_2SO_4 is first dissolved in (a) dilute sulfuric acid; (b) concentrated sulfuric acid; (c) water; (d) carbon disulfide. ______ 7
8. The initial dissolving of sulfur trioxide during the industrial production of H_2SO_4 produces (a) sulfuric acid; (b) sulfurous acid; (c) pyrosulfuric acid; (d) hydrosulfuric acid. ______ 8
9. Concentrated sulfuric acid is (a) watery and dense; (b) watery and not dense; (c) oily and dense; (d) oily and not dense. ______ 9
10. Concentrated sulfuric acid contains roughly what percentage of water? (a) 2% (b) 12% (c) 32% (d) 52% ______ 10
11. The making of which of the following does NOT generally involve the use of sulfuric acid? (a) nitroglycerin (b) paints (c) plastics (d) leather goods ______ 11

DIRECTIONS: In the parentheses at the right of each substance in the first column, write the letter of the formula in the second column that is most closely related.

12. sulfurous acid ( )
13. sulfur dioxide ( )
14. pyrosulfuric acid ( )
15. sulfuric acid ( )

a. S_2O
b. S_2O_2
c. SO_2
d. H_2S
e. H_2SO_3
f. H_2SO_4
g. $H_2S_2O_7$

Name ______________________ Class __________ Date __________

The Halogen Family

Section Review 28.1

DIRECTIONS: Write on the line at the right of each statement the letter preceding the word or expression that best completes the statement.

1. The halogen family makes up which group in the Periodic Table? (a) 14 (b) 15 (c) 16 (d) 17 ______ 1
2. The name "halogen" comes from the Greek words meaning (a) "salt-producer;" (b) "acid-former;" (c) "bad odor;" (d) "poisonous substance." ______ 2
3. Which of the following is a halogen? (a) carbon (b) oxygen (c) iodine (d) barium ______ 3
4. Which of the following is NOT a halogen? (a) bromine (b) astatine (c) fluorine (d) tellurium ______ 4
5. A halogen atom that has attained a complete set of outermost electrons has become a(n) (a) positive halogen ion; (b) positive halide ion; (c) negative halide ion; (d) electrolytic ion. ______ 5
6. In the elemental state, halogens exist as (a) individual atoms; (b) covalent diatomic molecules; (c) covalent triatomic molecules; (d) polyatomic ions. ______ 6
7. Going down the halogen column in the Periodic Table, melting and boiling points (a) decrease; (b) increase; (c) remain essentially the same; (d) do not follow any trend. ______ 7
8. Fluorine can be prepared from its compounds (a) only by chemical reduction; (b) only by chemical oxidation; (c) only by electrolysis; (d) by both chemical reduction and electrolysis. ______ 8
9. Compared to the other hydrogen halides, hydrogen fluoride is (a) much less polar; (b) slightly less polar; (c) of essentially equal polarity; (d) more polar. ______ 9
10. Hydrogen fluoride molecules associate mainly by (a) London forces; (b) van der Waals forces; (c) hydrogen bonding; (d) ionic attraction. ______ 10
11. In terms of its ability to act as an oxidizing agent, fluorine, as compared to the other halogens, is (a) better; (b) essentially equal; (c) slightly worse; (d) much worse. ______ 11

DIRECTIONS: Write on the line at the right of each statement the word or expression that best completes the meaning when substituted for the corresponding number.

12. A halogen atom has __(12)__ valence electrons. __________ 12
13. The halogen named __(13)__ is a liquid at room temperature. __________ 13
14. The halogen __(14)__ has the highest electronegativity. __________ 14
15. __(15)__ is the only halogen that forms a hydrogen halide that is a weak acid. __________ 15

Name ______________________ Class ____________ Date ____________

Fluorine

Section Review **28.2**

DIRECTIONS: Write on the line at the right of each statement the letter preceding the word or expression that best completes the statement.

1. How frequently does fluorine occur free in nature? (a) always (b) usually (c) rarely (d) never ______ 1
2. Which of the following is the formula of fluorspar? (a) CaF_2 (b) Na_3AlF_6 (c) FeF_2 (d) $Ca_5(PO_4)_3F$ ______ 2
3. Which of the following is the formula of fluorapatite? (a) CaF_2 (b) Na_3AlF_6 (c) FeF_2 (d) $Ca_5(PO_4)_3F$ ______ 3
4. Which of the following is the formula of cryolite? (a) CaF_2 (b) Na_3AlF_6 (c) FeF_2 (d) $Ca_5(PO_4)_3F$ ______ 4
5. Fluorine is commonly prepared by (a) replacement from its compounds by other halogens; (b) cyclotron processes; (c) electrolysis of KF and HF; (d) electrolysis of CF_4. ______ 5
6. When fluorine is prepared by electrolysis, (a) carbon acts as the positive electrode; (b) carbon acts as the negative electrode; (c) iron acts as the positive electrode; (d) iron acts as the negative electrode. ______ 6
7. Fluorine prepared by electrolysis does not react with metal in the cell because (a) the fluorine is unreactive; (b) the metal is unreactive; (c) a protective fluorocarbon coating is formed; (d) a protective metal fluoride coating is formed. ______ 7
8. Which of the following reacts only slowly with fluorine? (a) gold (b) neon (c) sodium (d) calcium ______ 8
9. Containers used to transport fluorine are usually made of (a) platinum; (b) carbon steel; (c) argon; (d) lead. ______ 9
10. Which of the following is used to etch glass? (a) dichlorodifluoromethane (b) hydrofluoric acid (c) cryolite (d) sodium fluoride ______ 10
11. Which of the following is used as a solvent for aluminum oxide in aluminum production? (a) dichlorodifluoromethane (b) hydrofluoric acid (c) cryolite (d) sodium fluoride ______ 11
12. Fluorides of noble gases (a) are completely unreactive; (b) are slightly reactive; (c) are moderately to highly reactive; (d) do not exist. ______ 12

DIRECTIONS: Complete the following statements, forming accurate sentences.

13. The only chemical family that contains elements that do not form compounds with fluorine is the ______________________________. 13
14. A fluorine compound that is used as a catalyst in producing high-octane gasoline is ______________________________. 14
15. A Freon that is commonly used as a refrigerant is ______________________________. 15

Chlorine

Section Review 28.3

DIRECTIONS: Write on the line at the right of each statement the letter preceding the word or expression that best completes the statement.

1. How frequently does chlorine occur free in nature? (a) always (b) usually (c) rarely (d) never _______ 1
2. Chlorides are found naturally in abundance in all of the following EXCEPT (a) cesium; (b) sodium; (c) magnesium; (d) potassium. _______ 2
3. Sodium chloride is generally found in large amounts in all of the following EXCEPT (a) sea water; (b) volcanic gases; (c) rock salt deposits; (d) underground brines. _______ 3
4. Elemental chlorine was first isolated by (a) Scheele; (b) Lavoisier; (c) Dalton; (d) Mendeleev. _______ 4
5. Preparation of chlorine from hydrogen chloride generally involves application of (a) heat only; (b) electricity only; (c) wind only; (d) both heat and electricity. _______ 5
6. Chlorine is released if hydrochloric acid is added to (a) sodium chloride; (b) calcium chloride; (c) calcium chlorate; (d) calcium hypochlorite. _______ 6
7. Hydrogen and chlorine react rapidly (a) in the dark only; (b) in sunlight only; (c) in black light only; (d) both in the dark and in sunlight. _______ 7
8. Chlorine combines with water to form (a) $HClO$ and HCl; (b) $HClO_2$ and HCl; (c) Cl_2O and HCl; (d) Cl_2O and H_2. _______ 8
9. Hydrogen chloride is often prepared by treating (a) aqueous chlorine with hydrogen peroxide; (b) sodium chloride with sulfuric acid; (c) sodium chloride with hydrogen peroxide; (d) sodium hypochlorite with nitric acid. _______ 9
10. Hydrogen chloride, in its most important commercial source, is formed as a byproduct of (a) electrolysis; (b) desalination; (c) chlorination; (d) buffering. _______ 10
11. Hydrochloric acid generally reacts with (a) no metals; (b) only metals above hydrogen in the activity series; (c) only metals below hydrogen in the activity series; (d) all metals. _______ 11

DIRECTIONS: In the parentheses at the right of each expression in the first column, write the letter of the substance in the second column that is most closely related.

12. often used as a catalyst ( )
13. used to make plastic objects ( )
14. often used as a solvent ( )
15. present in significant quantities in gastric juice ( )

a. polyvinyl chloride
b. carbon tetrachloride
c. aluminum chloride
d. potassium chloride
e. hydrochloric acid

Name ______________________ Class __________ Date __________

Bromine

Section Review **28.4**

DIRECTIONS: Write on the line at the right of each statement the letter preceding the word or expression that best completes the statement.

1. Bromine naturally forms commonly occurring salts with all of the following EXCEPT (a) magnesium; (b) barium; (c) potassium; (d) sodium. ______ 1
2. Bromine salts are commonly found in all of the following EXCEPT (a) lake water; (b) sea water; (c) underground brines; (d) salt deposits. ______ 2
3. Bromine is often prepared in the laboratory by combining sulfuric acid, sodium bromide, and (a) sulfur dioxide; (b) carbon dioxide; (c) lead dioxide; (d) manganese dioxide. ______ 3
4. Bromine has a disagreeable odor and is (a) more dense than water; (b) about the same density as water; (c) less dense than water; (d) the same color as water. ______ 4
5. Bromine is harmful in (a) vapor phase only; (b) liquid phase only; (c) both vapor and liquid phases; (d) gas phase only. ______ 5
6. Aqueous bromine solution is called (a) hydrobromic acid; (b) bromic acid; (c) bromine water; (d) bromine hydrate. ______ 6
7. Bromine and hydrogen (a) do not react with one another; (b) react to form hydrogen bromide; (c) react to form bromine hydride; (d) react to form hydrogen bromite. ______ 7
8. Aqueous bromine is a (a) weak oxidizing agent; (b) strong oxidizing agent; (c) weak reducing agent; (d) strong reducing agent. ______ 8
9. Bromine-containing organic compounds are added to polyurethane to make it (a) softer; (b) harder; (c) flame-retardant; (d) less soluble. ______ 9
10. Methyl bromide is used (a) as a soil fumigant; (b) in making photographic film; (c) in antiknock gasolines; (d) in oil-well drilling. ______ 10
11. Bromides of sodium and potassium are used frequently as (a) disinfectants; (b) stimulants; (c) pain-killers; (d) sedatives. ______ 11

DIRECTIONS: In the parentheses at the right of each expression in the first column, write the letter of the substance in the second column that is most closely related.

12. soil fumigant	( )	a. carbon tetrabromide
13. used in oil-well drilling	( )	b. silver bromide
14. used in making photographic film	( )	c. calcium bromide
		d. ethylene bromide
15. used in antiknock gasolines	( )	e. methyl bromide

Name ______________________ Class __________ Date __________

Iodine

Section Review 28.5

DIRECTIONS: Write on the line at the right of each statement the letter preceding the word or expression that best completes the statement.

1. Iodine occurs mainly in nature as (a) the free element; (b) iodides; (c) iodites; (d) iodates and periodates. ______ 1
2. Iodine occurs as an impurity mainly in deposits of (a) sodium chloride; (b) sodium bromate; (c) sodium nitrate; (d) sodium sulfate. ______ 2
3. The largest source of iodine in the United States is (a) sea water; (b) underground brine; (c) salt deposits; (d) volcanic rock. ______ 3
4. Most iodine in the United States is found in the state of (a) Oregon; (b) Texas; (c) New Mexico; (d) California. ______ 4
5. In the usual laboratory preparation, iodine initially appears as a (a) solute; (b) liquid; (c) vapor; (d) solid. ______ 5
6. Iodine forms compounds with all of the following EXCEPT (a) oxygen; (b) nitrogen; (c) sulfur; (d) carbon. ______ 6
7. In which of the following is iodine LEAST soluble? (a) pure water (b) aqueous iodide solutions (c) alcohol (d) carbon disulfide ______ 7
8. Carbon-iodide bonds (a) cannot form; (b) are weak; (c) are moderately strong; (d) are very strong. ______ 8
9. Detergents to which iodine is complexed (a) become completely ineffective; (b) become more soluble; (c) sanitize but do not serve as cleaning agents; (d) serve as both sanitizing and cleaning agents. ______ 9
10. Which of the following is commonly used as an antiseptic on the skin? (a) iodine complexed with organic compounds (b) silver iodide (c) potassium iodide (d) sodium iodide ______ 10
11. Which of the following is used for its property of light-sensitivity? (a) iodine complexed with organic compounds (b) silver iodide (c) potassium iodide (d) sodium iodide ______ 11

DIRECTIONS: Complete the following statements, forming accurate sentences.

12. Iodine vapor is colored ______________________. 12
13. Iodine-alcohol solution is colored ______________________. 13
14. An iodine compound that is commonly added to table salt is ______________________. 14
15. An iodine compound that is commonly used in photographic film is ______________________. 15

Name ________________ Class ________ Date ________

The Composition and Structure of the Nucleus

Section Review 29.1

DIRECTIONS: Write on the line at the right of each statement the letter preceding the word or expression that best completes the statement.

1. The force that keeps nucleons together is (a) the strong force; (b) the weak force; (c) electromagnetic force; (d) gravity. ______ 1
2. The approximate size of the nucleus is (a) 10^{-3} cm; (b) 10^{-6} cm; (c) 18^{-8} cm; (d) 10^{-13} cm. ______ 2
3. Which of the following are significantly affected by conditions of temperature, pressure, or catalysts? (a) both chemical and nuclear reactions (b) neither chemical nor nuclear reactions (c) chemical reactions only (d) nuclear reactions only ______ 3
4. Which of the following correctly relates mass and energy? (a) $E = mc^2$ (b) $E = mc$ (c) $E^2 = mc$ (d) $E = m^2c$ ______ 4
5. Which of the following have relatively small binding energies per nuclear particle? (a) light elements only (b) elements of intermediate mass (c) heavy elements only (d) both light and heavy elements ______ 5
6. The isotope whose atomic mass is defined as being exactly 12 *u* is (a) ${}^{6}_{6}C$; (b) ${}^{12}_{7}N$; (c) ${}^{12}_{5}B$; (d) ${}^{12}_{6}C$. ______ 6
7. The stability of atomic nuclei is most affected by the (a) number of neutrons; (b) number of protons; (c) number of electrons; (d) ratio of neutrons to protons. ______ 7
8. The presence of neutrons in high nuclear energy levels rather than in the lowest levels tends to (a) lower stability; (b) increase stability; (c) have no effect on stability, but to increase binding energy; (d) have no effect on either stability or binding energy. ______ 8
9. When a nucleus releases an alpha particle, it undergoes (a) nuclear disintegration; (b) fusion; (c) radioactive decay; (d) fission. ______ 9
10. When a nucleus releases a neutron, it undergoes (a) nuclear disintegration; (b) fusion; (c) radioactive decay; (d) fission. ______ 10
11. Which of the following tends to produce nuclei of lower mass? (a) fission only (b) fusion only (c) both fission and fusion (d) neither fission nor fusion ______ 11

DIRECTIONS: Write on the line at the right of each statement the word or expression that best completes the meaning when substituted for the corresponding number.

12. Any reaction in which the center of an atom is changed is called a(n) __(12)__ reaction. ________ 12
13. The equation __(13)__ represents the correct relationship between E, m, and c. ________ 13
14. For atoms of high atomic number, the most stable nuclei are those that have a neutron to proton ratio of roughly __(14)__ to 1. ________ 14
15. __(15)__ is the splitting of a heavy nucleus to form nuclei of intermediate mass. ________ 15

Name ______________________ Class __________ Date __________

The Phenomenon of Radioactivity

Section Review 29.2

DIRECTIONS: Write on the line at the right of each statement the letter preceding the word or expression that best completes the statement.

1. Compared to uranium, radium is (a) much less radioactive; (b) slightly less radioactive; (c) slightly more radioactive; (d) much more radioactive. ______ 1
2. Radioactive nuclides cause air to (a) become ionized; (b) fluoresce; (c) condense; (d) become radioactive. ______ 2
3. Half-life is the length of time in which half of the atoms in a sample (a) undergo radioactive decay; (b) undergo nuclear fission; (c) undergo nuclear fusion; (d) react chemically. ______ 3
4. The half-life of radium-226 is roughly (a) 16 seconds; (b) 16 hours; (c) 16 years; (d) 1600 years. ______ 4
5. Which of the following are deflected most by a magnetic field? (a) alpha particles (b) beta particles (c) gamma rays (d) None of the above are deflected at all. ______ 5
6. Which of the following generally have the lowest penetrating ability? (a) alpha particles (b) beta particles (c) gamma rays (d) All of the above have the same penetrating ability. ______ 6
7. Complete the following equation. $^{238}_{92}U$ + ______ → $^{239}_{92}U$ (a) $^{4}_{2}He$ (b) $^{0}_{1}n$ (c) $^{1}_{1}H$ (d) $^{0}_{-1}e$ ______ 7
8. Complete the following equation. $^{239}_{93}Np$ → ______ + $^{0}_{-1}e$ (a) $^{239}_{90}U$ (b) $^{239}_{92}U$ (c) $^{239}_{94}Pu$ (d) $^{238}_{94}Pu$ ______ 8
9. Which of the following is generally an induced radioactive nuclide? (a) $^{238}_{92}U$ (b) $^{60}_{27}Co$ (c) $^{235}_{92}U$ (d) $^{17}_{8}O$ ______ 9

DIRECTIONS: In the parentheses at the right of each word or expression in the first column, write the letter of the nuclear symbol in the second column that is most closely related.

10. beta particle ( ) a. $^{1}_{1}H$
11. proton ( ) b. $^{4}_{2}He$
12. neutron ( ) c. $^{2}_{1}H$
13. alpha particle ( ) d. $^{1}_{0}n$

e. $^{0}_{-1}e$

DIRECTIONS: Write the answer to question 14 on the line to the right, and show your work in the space provided.

14. Calculate the half-life of an isotope if 125 g of a 500 g sample of the isotope remains after 3.0 years. ______ 14

Name ______________________ Class __________ Date __________

Applications of Radioactivity

Section Review 29.3

DIRECTIONS: Write on the line at the right of each statement the letter preceding the word or expression that best completes the statement.

1. The half-life of carbon-14 is about (a) 57 years; (b) 570 years; (c) 5700 years; (d) 57 000 years. ______ 1
2. The isotope $^{14}_{6}C$ decays into (a) $^{14}_{7}N$; (b) $^{15}_{7}N$; (c) $^{10}_{4}Be$; (d) $^{18}_{8}O$. ______ 2
3. Radioactive dating has revealed that moon rocks are roughly (a) 4 million years old; (b) 40 million years old; (c) 400 million years old; (d) 4 billion years old. ______ 3
4. Radioactive dating is most directly based upon a knowledge of a substance's (a) melting point; (b) half-life; (c) rate of weathering; (d) heat of reaction. ______ 4
5. The isotope $^{40}_{19}K$ decays into (a) $^{14}_{6}C$; (b) $^{36}_{17}Cl$; (c) $^{40}_{18}Ar$; (d) $^{41}_{19}Ar$. ______ 5
6. The isotope $^{60}_{27}Co$ is commonly used for all of the following EXCEPT to (a) kill food-spoilage bacteria; (b) preserve food; (c) kill insects that infest food; (d) treat heart disease. ______ 6
7. Radioactive drugs are used for (a) diagnostic purposes only; (b) blood and tissue testing only; (c) both diagnostic and blood and tissue testing purposes; (d) pain killers. ______ 7
8. Radioactive tracers are used to obtain information on the movement of which of the following along sea coasts? (a) water (b) salts (c) sand (d) fish ______ 8
9. Radioactive tracers are used to obtain information on what aspect of fertilizers? (a) their efficiency (b) their purity (c) their chemical composition (d) their natural radioactivity ______ 9
10. Many new radioactive nuclides are presently made in a nuclear reactor at (a) Oak Ridge, Tennessee; (b) Atlanta, Georgia; (c) Savanna, Georgia; (d) Nashville, Tennessee. ______ 10

DIRECTIONS: Complete the following statements, forming accurate sentences.

11. The age of once-living materials is generally determined by measuring their level of the isotope ______________________. 11
12. The oldest known minerals on the earth were found on the continent of ______________________. 12
13. Radioactive carbon is naturally produced in the atmosphere by the action of ______________________. 13
14. A radioactive isotope commonly used to treat cancer is an isotope of the element ______________________. 14
15. Radioactive materials whose movement is used to monitor processes and measure flow rates are called ______________________. 15

Name ______________________ Class __________ Date __________

Energy from the Nucleus

Section Review 29.4

DIRECTIONS: Write on the line at the right of each statement the letter preceding the word or expression that best completes the statement.

1. Nuclear fission tends to produce (a) alpha particles; (b) neutrons and lower-mass nuclei; (c) neutrons only; (d) neutrons and higher-mass nuclei. ______ 1
2. Fission is basically a process of nuclear (a) combination; (b) decay; (c) splitting; (d) disintegration. ______ 2
3. After uranium-235 captures a slow neutron, which of the following occurs? (a) nothing (b) fusion (c) fission (d) production of a transuranium nuclide ______ 3
4. The energy released during nuclear fission comes from (a) stored chemical energy; (b) the surroundings; (c) electromagnetic energy; (d) mass-to-energy conversion. ______ 4
5. Fusion is basically a process of nuclear (a) combination; (b) decay; (c) splitting; (d) disintegration. ______ 5
6. Which of the following provides energy in the sun? (a) chemical combustion (b) nuclear fusion (c) nuclear fission (d) electrolytic reduction ______ 6
7. In the earliest nuclear reactors, the moderator was typically (a) uranium; (b) boron steel; (c) graphite; (d) water. ______ 7
8. Pressurized water in modern nuclear power plants generally serves as (a) both a moderator and a coolant; (b) a coolant only; (c) a moderator only; (d) a heat source. ______ 8
9. The heat produced by a reactor is used to produce (a) steam; (b) molten metal; (c) graphite; (d) coal. ______ 9
10. Ionized fuel in fusion reactions is confined (a) in steel containers; (b) in lead containers; (c) in water-cooled plastic containers; (d) by magnetic fields. ______ 10
11. The quantity of uranium required to sustain a chain reaction is called the (a) threshold mass; (b) critical volume; (c) critical mass; (d) Curie mass. ______ 11

DIRECTIONS: In the parentheses at the right of each expression in the first column, write the letter of the expression in the second column that is most closely related.

12. limits the number of free neutrons	( )	a. moderator
13. used to slow neutrons	( )	b. coolant
14. drives an electric generator	( )	c. control rod
15. produces new radioactive substances and energy by fission	( )	d. nuclear reactor
		e. turbine